Automated Optimization-Based Synthesis of Distributed Energy Supply Systems

Von der Fakultät für Maschinenwesen der
Rheinisch-Westfälischen Technischen Hochschule Aachen
zur Erlangung des akademischen Grades eines Doktors
der Ingenieurwissenschaften genehmigte Dissertation

vorgelegt von

Philip Voll

Berichter: Univ.-Prof. Dr.-Ing. André Bardow
Prof. Nilay Shah, PhD

Tag der mündlichen Prüfung: 28. Juni 2013

Diese Dissertation ist auf den Internetseiten der Hochschulbibliothek verfügbar.

Aachener Beiträge zur Technischen Thermodynamik Band 1
Philip Voll
Automated Optimization-Based Synthesis of Distributed Energy Supply Systems

ISBN: 978-3-86130-474-6

Bibliografische Information der Deutschen Bibliothek
Die Deutsche Bibliothek verzeichnet diese Publikation in der Deutschen Nationalbibliografie; detaillierte bibliografische Daten sind im Internet über http://dnb.ddb.de abrufbar.

Vertrieb:

1. Auflage 2014
© Wissenschaftsverlag Mainz GmbH - Aachen
Süsterfeldstr. 83, 52072 Aachen
Tel. 0241/87 34 34
Fax 0241/87 55 77
www.Verlag-Mainz.de

Herstellung:

Druck und Verlagshaus Mainz GmbH Aachen
Süsterfeldstraße 83
52072 Aachen
Tel. 0241/87 34 34
Fax 0241/87 55 77
www.DruckereiMainz.de

Satz: nach Druckvorlage des Autors
Umschlaggestaltung: Druckerei Mainz

printed in Germany
D82 (Diss. RWTH Aachen University, 2013)

Vorwort

Die vorliegende Arbeit entstand im Rahmen meiner Tätigkeit als wissenschaftlicher Mitarbeiter am Lehrstuhl für Technische Thermodynamik der Rheinisch-Westfälischen Technischen Hochschule Aachen.

Mein größter Dank gilt meinem Doktorvater, Herrn Prof. Dr.-Ing. André Bardow, der mich in meinem Promotionsvorhaben stets unterstützt hat. Ich bedanke mich herzlich für die anregenden Diskussionen und die gestalterischen Freiräume, die wesentlich zum Gelingen dieser Arbeit beigetragen haben. Weiterhin danke ich Herrn Prof. Dr. Nilay Shah (Imperial College London) für die fachlichen Diskussionen und die Übernahme des Koreferats. Herrn Prof. Dr.-Ing. Dirk Müller danke ich für die Übernahme des Prüfungsvorsitzes und die angenehme Art, die Prüfung zu leiten.

Mein herzlicher Dank gilt allen gegenwärtigen und ehemaligen Mitarbeitern des Lehrstuhls für Technische Thermodynamik, die mich während meiner Forschungstätigkeit unterstützt oder durch die sehr gute Arbeitsatmosphäre motiviert haben – vor allem den Mitgliedern der Energiesystemtechnik Arbeitsgruppe. Besonders hervorheben möchte ich Stefan Kirschbaum, der mir sehr bei der Vorbereitung meines Promotionsthemas geholfen hat; Matthias Lampe, der viele Stunden seines Lebens der Diskussion meiner Arbeit geopfert hat; und Carsten Klaffke, der meine Ideen implementiert hat. Schließlich danke ich meinem langjährigen Weggefährten, Johannes Jung, für die tolle gemeinsame Zeit im *Office*.

Des Weiteren danke ich allen Studenten, die zum Gelingen dieser Arbeit beigetragen haben. Ich danke vor allem Maike Hennen, Katharina Wolters und Stefan Schwering.

Außerdem danke ich meinen Eltern für ihre liebevolle Unterstützung. Abschließend danke ich meiner Frau, Anna, die mir stets mit Rat und Tat zur Seite stand, gerade in Zeiten als mein Doktorvater den Lehrstuhl noch nicht geleitet hat. In den letzten Zügen meiner Promotion hast du mir den Rücken frei gehalten – ich verstehe erst jetzt wie viel dir das abverlangt hat. Danke für alles!

Aachen, im August 2013 *Philip Voll*

'Le mieux est l'ennemi du bien.'

(The best is the enemy of the good.)

Voltaire, *La Béqueule* (1772)

Contents

List of Figures

List of Tables

Notation

Latin symbols

$C_{t_{\mathrm{CF}}}$	net present value at cash flow time t_{CF}	€
d	decision variable vector representing decisions on the design level	-
D	design decision variable space	-
$\mathbf{C}$	connectivity matrix	-
$\dot{E}$	energy flow	W
i	discount rate	-
I	investment	€
M	constant coefficient in cost function	-
n	number (general counter symbol)	-
o	decision variable vector representing decisions on the operation level	-
O	operation decision variable space	-
p_U	price of input energy	€/kWh
P	probability, weighting factor	-
q	load fraction $\dot{Q}/\dot{Q}_{\mathrm{N}}$	-
$\dot{Q}$	thermal power, heat flux	W
$R_{t_{\mathrm{CF}}}$	net cash flow at cash flow time t_{CF}	€
s	decision variable vector representing decisions on the synthesis level	-
S	synthesis decision variable space	-
t	time	s
u	relative input power	-
$\dot{U}_{nt}$	input power	W
v	relative output power (load fraction)	-
$\dot{V}$	output power	W
y	binary decision variable representing the (non-)existence of a conversion unit	-

Greek symbols

γ	binary decision variable representing the choice of a piece of linearized cost function	-
δ	binary decision variable representing the state of operation of a conversion unit	-
η	efficiency	-
σ	an energy supply structure evolved by mutation	-
Σ	set of all possible energy supply structures	-
Ψ	continuous auxiliary decision variable	W
ξ	continuous auxiliary decision variable	W

Subscripts and superscripts

0	cooling
b	binary decision variable
B	base, base case
c	constraint
CF	cash flow
el	electric
n	counter of energy conversion units n
N	nominal
t	time, counter of time steps t
th	thermal
v	continuous decision variable
Δ	change

Abbreviations

AC	absorption chiller
B	boiler
CG	cooling generator
CED	cumulative energy demand
CHP	combined heat and power
COP	coefficient of performance
DESS	distributed energy supply system(s)
EA	evolutionary algorithm

EG	electricity generator
ES	evolution strategy
EU	energy user
FS	fuel supplier
GA	genetic algorithm
HD	heat demand
HG	heat generator
HU	heat energy user
IP	integer program
LP	linear program
MILP	mixed-integer linear program
MINLP	mixed-integer nonlinear program
NLP	nonlinear program
NGH	natural gas hook-up
NPV	net present value
PI	power input
TC	turbo-driven compression chiller (turbo-chiller)

Miscellaneous symbols

$\|\|$	parallel, parallel connection
$\{\}$	delete, deletion

Kurzfassung

Der konzeptionelle Entwurf dezentraler Energieversorgungssysteme stellt ein komplexes Problem dar, das durch eine unüberschaubare Vielfalt möglicher Kombinationen der verschiedenen Versorgungsanlagen inklusive deren Auslegung und Betrieb charakterisiert ist. In der Regel führen Ingenieure zur Lösung dieses Problems Variantenvergleiche mittels Simulationsstudien durch. Dieser Ansatz erfordert allerdings die explizite Vorgabe aller zu bewertenden Alternativen, wodurch praktisch nur eine geringe Anzahl an Versorgungskonzepten untersucht werden kann. Insbesondere kann nicht garantiert werden, dass die optimale Lösung identifiziert wird. Der Einsatz mathematischer Optimierungsmethoden ermöglicht hingegen die Berücksichtigung einer unbegrenzten Anzahl von Varianten, um vorurteilsfrei das optimale Energieversorgungssystem zu identifizieren.

Für den optimierungsbasierten Entwurf von Energieversorgungssystemen wird üblicherweise die Methodik der *Superstruktur-Optimierung* eingesetzt. Diese Methodik ermöglicht die Bestimmung der optimalen Versorgungsstruktur inklusive Dimensionierung und Betrieb der installierten Anlagen. Allerdings muss eine Superstruktur vorgegeben werden, die alle denkbaren Lösungsstrukturen enthält. Dieser Schritt stellt wiederum ein komplexes Problem dar: Zum Einen kann nicht gewährleistet werden, dass die Superstruktur die optimale Lösung enthält. Zum Anderen ist die Optimierung einer unnötig großen Superstruktur mit nicht vertretbarem Rechenaufwand verbunden. In der Literatur ist derzeit keine Methode verfügbar, die in geeigneter Weise den Entwurf dezentraler Energieversorgungssysteme ohne Vorgabe einer Superstruktur ermöglicht.

In der vorliegenden Arbeit werden zwei neue Methoden für den optimierungsbasierten Entwurf dezentraler Energieversorgungssysteme vorgestellt, die den Entwurfsprozess automatisieren, um die Vorgabe einer Superstruktur durch den Nutzer zu vermeiden. Es handelt sich um eine *superstrukturbasierte* und eine *superstrukturfreie* Methodik. Die superstrukturbasierte Methodik wendet einen Algorithmus zur automatischen Superstruktur-Generierung an. Die zu Grunde liegende Optimierungsstrategie erhöht ausgehend von einer minimalen Superstruktur sukzessive die Zahl der in der Superstruktur vorgesehenen Anlagen, bis eine erneute Optimierung keine bessere Lösung mehr identifiziert – also die optimale Lösung gefunden ist. Für den superstrukturfreien Ansatz wird ein wissensintegrierter Evolutionärer Algorithmus eingesetzt, der mittels Ersetzungsregeln Teile einer Lösung gegen alternative Teilstrukturen ersetzt, um so neue Lösungen zu generieren, die anschließend hinsichtlich der Anlagendimensionierung und des Anlagenbetriebs optimiert werden. Zur automatischen Definition der Ersetzungsregeln werden alle betrachteten Versorgungstechnologien in einem hierarchisch strukturierten Graphen, der sogenannten *Energy Conversion Hierarchy*, hinsichtlich ihrer Funktionen klassifiziert. Beide Methoden beruhen auf einer generischen, komponentenbasierten Modellierung, die den Einsatz beliebiger Modellformulierungen ermöglicht. In der vorliegenden Arbeit wird quasistationäres Systemverhalten angenommen und eine robuste gemischt-ganzzahlige lineare (MILP) Formulierung gewählt. Die MILP Formulierung berücksichtigt eine kontinuierliche Anlagendimensionierung, Größenabhängigkeit der Investitionskosten sowie die Betriebscharakteristika der Versorgungstechnologien.

Der optimierungsbasierte Entwurf mittels der neu entwickelten Methoden wird anhand eines Beispiels der Pharmaindustrie demonstriert. Das betrachtete Unternehmen besitzt zeitlich variable Energiebedarfe, die auf dem Produktionsstandort verteilt vorliegen. Als Zielfunktion zur Optimierung wird der Kapitalwert maximiert. Der optimierungsbasierte Entwurf führt zu einer kostengünstigen und energieeffizienten Lösung, die die bestehende Infrastruktur, d.h. bereits installierte Anlagen sowie bauliche Einschränkungen, berücksichtigt. Zusätzlich werden nahoptimale Lösungsalternativen und Pareto-optimale Kompromisslösungen generiert. Die nahoptimalen Lösungsalternativen unterscheiden sich zwar hinsichtlich der installierten Anlagen, besitzen aber praktisch gleich gute Kapitalwerte. Die Pareto-optimalen Lösungen stellen die besten Kompromisslösungen einer zweikriteriellen Optimierung dar, die als Ziel die gleichzeitige Minimierung der Investitionen und des kumulierten Energieaufwands verfolgt. Abschließend werden die generierten Lösungen hinsichtlich weiterer Randbedingungen diskutiert, die während der Optimierung vernachlässigt wurden oder die sich im Laufe der Zeit ändern könnten. Somit wird der Entscheider in die Lage versetzt, die für den jeweiligen Anwendungsfall am besten geeignete Lösung auszuwählen. Das reale Fallbeispiel zeigt, dass die entwickelten Methoden einen automatisierten Entwurf optimaler Energieversorgungssysteme ermöglichen und somit entscheidend zur Akzeptanz von Optimierungswerkzeugen in der Praxis beitragen können.

Chapter 1

Introduction

Today, sustainable development is widely recognized as one of the most pressing challenges facing our world's societies. Sustainable development links environmental, economic, and social issues (Needham, 2011). The goal of sustainable development is usually defined as

> *"to meet the needs of the present without compromising the ability of future generations to meet their own needs"*. (United Nations, 1987)

A serious obstacle that needs to be overcome for sustainable development is the excessive use of fossil fuels (Wuebbles and Jain, 2001). In particular the associated greenhouse gas emissions (Solomon et al., 2007) and the widely predicted shortage of fossil fuels (Doman, 2011) call for urgent engineering solutions (Bakshi and Fiksel, 2003). To face these challenges, Rogner and Zhou (2007) identified the synthesis of energy-efficient energy systems as readily available solutions for the short- and the mid-term. Like sustainable development, the synthesis of energy-efficient energy systems links environmental and economic aspects; i.e., the synthesis does not only lead to primary energy savings and CO_2 emissions reductions, but it often also leads to significant cost savings as shown by various authors for building energy systems (Lozano et al., 2009; Liu et al., 2010a), district and urban energy systems (Bouvy and Lucas, 2007; Weber and Shah, 2011; Keirstead et al., 2012), and industrial energy systems (Aguilar et al., 2007a,b; Velasco-Garcia et al., 2011; Tina and Passarello, 2012).

During his time as Ph.D. student, the author of this thesis collaborated with several energy consultancies active in the field of industrial energy systems synthesis. The collaboration partners confirmed what Drumm et al. (2013) observed for the chemical industry: Energy costs in industry are usually of the same order of magnitude as the companies' profits, and thus the synthesis of energy-efficient energy systems is a key lever to increase the companies' profits. However, in industry, priority is given to production, and therefore any investments in energy-efficiency measures compete with investments in production facilities and product design. Therefore, thorough assessments are required for each investment to decide whether an energy-efficiency measure is cost effective or not. Today in engineering practice, these assessments mainly rely on simulation studies (Klatt and Marquardt, 2009). However, the simulation-based synthesis approach requires sequential and iterative decision making based on common sense, engineering experience, and intuition to manually define each and every synthesis alternative to be evaluated (Biegler et al., 1997). For this reason, the simulation-based synthesis approach limits the number of considered synthesis alternatives and generally results in suboptimal solutions only (Frangopoulos et al., 2002).

To systematize, improve, and automate the synthesis of energy systems, a vast amount of research has been performed in the past for the development of computer-aided, optimization-based synthesis methods (Connolly et al., 2010; Nakata et al., 2011). However, as Klatt and Marquardt (2009) discuss for the example of the chemical industry, "*synthesis methodologies relying on rigorous optimization are rarely used in industrial practice. This statement even holds for special cases such as heat exchanger design or distillation column sequencing and design but even more for the treatment of integrated processes.* (p. 540)"

As main issue for the lack of optimization-based synthesis methods in engineering practice, Grossmann (2012) points out that application of these methods usually involves many manual user inputs (e.g., for model and algorithm parameterization) that require detailed systems engineering knowledge, which, however, is often not prevailing among industrial practitioners (i.e., modeling skills, knowledge of optimization theory and solution algorithms, etc.).

In view of the described state-of-the-art of energy systems synthesis in engineering practice, it is the goal of this thesis to provide automated optimization-based methodologies for the synthesis of energy systems to make optimization accessible for practitioners.

1.1 Structure of this thesis

In chapter 2, an introduction is given to the synthesis of distributed energy supply systems (DESS). It is well understood that the decisions made during the conceptual synthesis phase fix a major part of the system's total cost (Biegler et al., 1997; Patel et al., 2009). Thus, this thesis focuses on the synthesis of distributed energy supply systems on the conceptual level. Next, a detailed review is given of optimization-based methods for the synthesis of DESS. Finally, based on the preceding review, drawbacks and shortcomings of available optimization-based synthesis methods are discussed, and the scope of this thesis is specified.

Next, in chapter 3, a mathematical programming framework is introduced for the modeling of distributed energy supply systems. For this purpose, a generic component-based modeling framework is presented. For each technology, a component model is defined that is composed of an investment cost curve and a performance model. The original mathematical programming formulation is a mixed-integer nonlinear programming (MINLP) formulation. For robust optimization, the MINLP formulation is linearized to derive a mixed-integer linear programming (MILP) formulation. Finally, the MILP formulation is evaluated with regard to its approximation of the original MINLP formulation and the required computational effort to solve a synthesis test problem.

In chapters 4 and 5, two novel methodologies are proposed for the automated optimization-based synthesis of distributed energy supply systems: a superstructure-based and a superstructure-free synthesis methodology. The superstructure-based synthesis methodology (chapter 4) automatically generates superstructures for DESS synthesis problems. The methodology continuously expands and optimizes these superstructures to identify the optimal solution. The superstructure-free synthesis methodology (chapter 5) combines evolutionary and deterministic algorithms to simultaneously generate and optimize solution alternatives to search for the optimal solution. Both methodologies are applied to an illustrative synthesis problem. For this purpose, they employ the generic MILP framework introduced in chapter 3. The methodologies are evaluated with regard to solution quality and computational performance.

In chapter 6, a real-world problem from the pharmaceutical industry is discussed. Both synthesis methodologies are applied to this problem. A synthesis procedure is presented that generates a set of promising solution alternatives instead of a single optimal solution only, thus providing deeper understanding of the synthesis problem at hand and opening up a wider space for rational synthesis decisions. The features of both methods are discussed with regard to practicality, and a comparative evaluation is provided.

Finally, the work is summarized and conclusions are drawn (chapter 7). Furthermore, topics for future research are identified. Last but not least, based on the experience gained during the work on this thesis, the author comments on the necessity of optimization-based methods for the synthesis of distributed energy supply systems.

Chapter 2

Optimization-based conceptual synthesis of distributed energy supply systems

In this chapter, a review is provided of optimization-based methods for the conceptual synthesis of distributed energy supply systems. First (section 2.1), an introduction is given to the systematic synthesis of distributed energy supply systems. Next, the fundamentals of mathematical optimization are briefly discussed (section 2.2). In section 2.3, a review is provided of optimization-based synthesis methods including a discussion of drawbacks and shortcomings of the available methods. In section 2.4, a brief introduction is given to multi-objective optimization methods in the field of energy systems synthesis. Finally, based on the given review, the contribution of this thesis is highlighted (section 2.5).

2.1 Systematic synthesis of distributed energy supply systems

2.1.1 Distributed energy supply systems (DESS)

Energy systems comprise two subsystems: the *energy supply system* and the *system of final energy users* (Fig. 2.1) (Augenstein et al., 2005). The energy supply system converts forms of secondary energy into final energy required by the final energy users. The final energy users employ the delivered energy in all kinds of technical processes converting it into useful energy and rejecting heat to the energy supply system. The final energy users correspond to a large variety of users such as chemical processes, manufacturing sites, or residential areas. The energy supply system is connected to the public energy market and the environment: The public energy market supplies various forms of secondary energy and takes in feed-in electricity from the supply system; the environment performs heat exchange with the examined site.

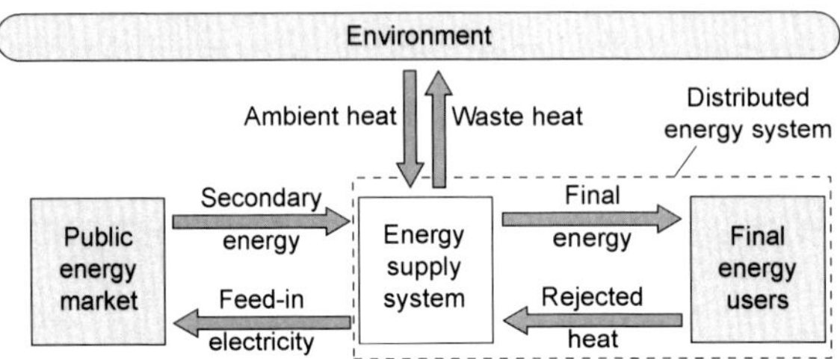

Figure 2.1: Energy flows between energy supply systems and neighboring systems (adapted from Augenstein et al. (2005)).

Energy supply systems are usually highly integrated and complex systems containing a multitude of technical components including energy conversion technologies, energy distribution infrastructure, and energy storage facilities. In an industrial context, energy supply systems are often referred to as *utility systems* (see, e.g., (Luo et al., 2011; Velasco-Garcia et al., 2011)). Conventionally, heat and power are generated in large units allowing for centralization, however, often at the expense of lower energy efficiency (Mehleri et al., 2012). In contrast, *distributed energy supply systems* (DESS) integrate centralized with distributed (also *decentralized*, or *on-site*) conversion technologies. Centralized conversion technologies usually have excellent economies of scale, but transfer the generated energy forms along long distances, and thus are typically subject to significant energy losses. Moreover, centralized conversion technologies cannot always be operated optimally with regard to the present energy demands. In contrast, the use of many smaller distributed energy conversion technologies enables to reduce the energy losses due to on-site power generation and optimal unit operation (Bouffard and Kirschen, 2008; Jiayi et al., 2008).

It should be noted that heat exchanger networks are usually assumed to be part of the final energy users, but not of the energy supply system. Thus, heat integration (Furman and Sahinidis, 2002; Morar and Agachi, 2010) is understood as a foregoing measure to reduce the demands of the final energy users, which then have to be met by the energy supply system.

2.1.2 Conceptual synthesis of distributed energy supply systems

In this work, the focus is on the conceptual synthesis of distributed energy supply systems. The conceptual synthesis phase usually fixes a major part of the system's total cost (Biegler et al., 1997; Patel et al., 2009), and thus it is of particular value to focus on the synthesis on the conceptual level. According to Mizsey and Fonyo (1990b), the conceptual synthesis task is to either create a new flowsheet (*grassroots* synthesis) or to improve an existing flowsheet (*retrofit* synthesis). It should be noted that retrofit synthesis always includes grassroots synthesis, and thus is the more complex task.

Frangopoulos et al. (2002) define the synthesis task as hierarchically-structured problem that has to be considered on three levels (Fig. 2.2):

- the synthesis level,
- the design level, and
- the operation level.

At the synthesis level, the engineer needs to optimize the structure or configuration of the energy system. This encompasses the selection of the technical components and the optimal layout of their interconnections. At the design level, one has to define the technical specifications (capacity, operating limits, etc.) of the units selected during synthesis. Finally, given the system synthesis and design, the optimal operation modes need to be specified for each instant of time at the operation level. The system design and operation directly influence the solution of the synthesis problem, and thus, for optimal systems synthesis, all three levels must be taken into account simultaneously.

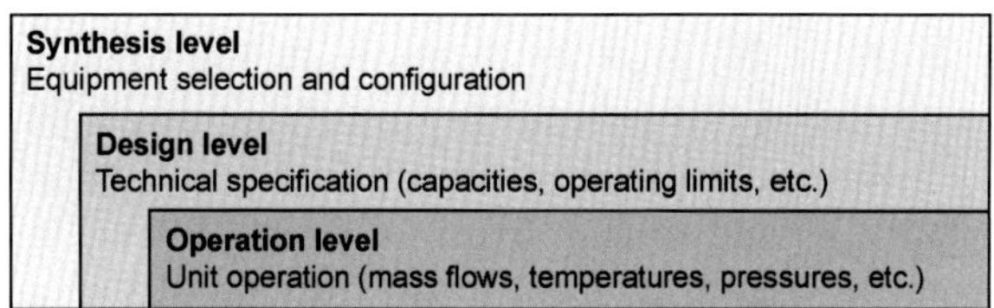

Figure 2.2: The synthesis task as hierarchically-structured problem on three levels.

Based on the conceptual synthesis, in the following detailed design stage, all significant details of the process are specified and engineering drawings and equipment lists are prepared.

For the conceptual synthesis of DESS, the special characteristics inherent to DESS have to be regarded, and therefore, the synthesis problem is inherently difficult to solve as a most simple example already shows: Consider the total cost optimization of a heating system for a single time-varying heat demand (Fig. 2.3 a). Even if only a single technology such as a simple boiler is regarded, the grassroots design problem becomes challenging due to the necessity to consider the following characteristics:

- economy of scale for equipment investments (Papoulias and Grossmann, 1983),
- limited capacities of standardized equipment (Yokoyama and Ito, 2006),
- equipment part-load performance (Velasco-Garcia et al., 2011), and
- minimum operation loads of the equipment (Prokopakis and Maroulis, 1996).

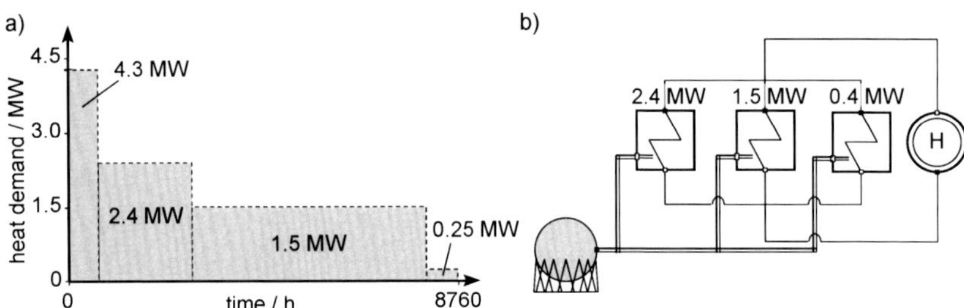

Figure 2.3: Motivating example for DESS synthesis problems. a) Time-varying heat demand. b) Optimal heating system.

On the one hand, the boilers' total capacities have to cover the maximum heat demand, and economy of scale of the investment cost favors large over small equipment; on the other hand, off-the-shelf boilers are available only within certain capacity ranges, and hence exceptionally large boilers become excessively expensive. Then again, the boilers' efficiencies drop at part-load operation, and thus equipment sizing enabling full-load operation is beneficial. Moreover, boilers must not be operated below their minimum part-loads of typically 20 %, and therefore need to be sized sufficiently small to cover minimum loads. For the considered example, accounting for the listed characteristics leads to the installation of three boilers in the optimal configuration (Fig. 2.3 b). This simple example shows that, in general, all of the listed characteristics need to be reflected for optimal DESS synthesis. Moreover, it is shown that multiple redundant units are generally to be expected in DESS. This is in stark contrast to classical process synthesis problems, for which multiplicity and redundancy are often regarded as shortcomings of the problem formulation (Farkas et al., 2005).

2.1.3 Systematic synthesis methods

In the past, a vast amount of research has been performed on the development of systematic methods for the synthesis of energy and process systems. The methodologies have been reviewed extensively by many authors (Frangopoulos et al., 2002; Li and Kraslawski, 2004; Westerberg, 2004; Barnicki and Siirola, 2004; Jaluria, 2008; Grossmann and Guillén-Gosálbez, 2010). Douglas (1985) proposed a hierarchical decision strategy for process synthesis, which serves as common basis for virtually any systematic synthesis method proposed thereafter until today. The hierarchical decision strategy exploits the systems nature of the synthesis problem to apply the *decomposition* and *aggregation* principles for systems analysis and synthesis, respectively. The synthesis methodologies proposed in literature are typically classified as

i) heuristic methods,

ii) targeting methods, and

iii) mathematical optimization methods.

Heuristic methods (i) use rules derived from engineering knowledge and experience to generate promising solutions. Thorough reviews are given by Mostow (1985), and Han and Stephanopoulos (1996). Targeting methods (ii) integrate physical insight to acquire a better understanding of the considered system. The most prominent representative of targeting methods is the exergy-based pinch methodology (Linnhoff et al., 1982) for heat integration. Chou and Shih (1987) and Sama et al. (1989) proposed exergy-based targeting methods for the system-wide synthesis of industrial energy supply systems.

To make these methods accessible for practitioners, in the past, a large variety of knowledge-based *expert systems* have been implemented for energy and process systems synthesis. To name just a few, for total process synthesis, there are AIDES (Siirola et al., 1971), BALTAZAR (Mahalec and Motard, 1977), the rule-based framework by Lu and Motard (1985), PIP (Kirkwood et al., 1988), and PROSYN[1] (Schembecker and Simmrock, 1997). More recently, process synthesis methods have been proposed based on the generic concept of *case-based reasoning* (CBR) (Pajula et al., 2001; Avramenko et al., 2005; Botar-Jid et al., 2010). For the synthesis of industrial energy supply systems, there is the expert system COLOMBO (Melli and Sciubba, 1997). Duic et al. (2008) proposed the *RenewIslands* methodology for the systematic generation of energy supply systems integrating renewable energy sources on islands. Expert systems can be very effective for generating promising candidate solutions

[1]not to be mistaken with the MINLP optimization framework PROSYN by Kravanja and Grossmann (1993)

(Kott et al., 1989; Sciubba and Melli, 1998). However, they generally identify only suboptimal solutions. To avoid this shortcoming, mathematical optimization methods have been developed.

Mathematical optimization methods (iii) (also *mathematical programming*, or simply *optimization*) enable the algorithmic solution of a synthesis problem formulated as mathematical model by searching for the solution that minimizes (or maximizes) the so-called *objective function*, i.e., the optimization criterion (Floudas, 1995). Hence, optimization considers all aspects represented in the problem formulation simultaneously, and therefore represents a much more comprehensive methodology than heuristic and targeting methods. It should be emphasized, however, that the integration of heuristic, targeting, and optimization methods can potentially improve the overall synthesis process, and thus the three methodologies should not be understood as competitive, but as complementary (Li and Kraslawski, 2004): The targets generated from the targeting methods can be employed as heuristics or rules in the heuristic methods, which in turn can be embedded in optimization procedures to the benefit of the whole synthesis process. As an example, Mizsey and Fonyo (1990a,b) implemented an interactive user-driven approach that integrates heuristic and optimization methods for the synthesis of total process flowsheets.

In the past, optimization has been applied exhaustively to the synthesis of both energy and process systems. In the remainder of this chapter, a review is given of optimization-based methods with special attention to the synthesis of energy systems.

2.2 Mathematical optimization

A constrained mathematical optimization problem is represented by a number of *equality* and *inequality constraints* (the model) that define the relations between the *decision variables*, whose values have to be determined to minimize (or maximize) an *objective function* (the optimization criterion). The mathematical notation "minimize the objective function subject to (s.t.) the constraints on its decision variables" is as follows:

$$\begin{aligned} \min_{x} \quad & f(x), && \text{"objective function"} \\ \text{s.t.} \quad & h(x) = 0, && \text{"equality constraints"} \\ & g(x) \leq 0, && \text{"inequality constraints"} \end{aligned} \tag{2.1}$$

where x is the vector of decision variables, f is the objective function to be minimized, and h and g are vector functions of the variables x describing the equality and inequality constraints. Depending on whether the objective function and the (in-)equality relations are linear or nonlinear, and if the problem incorporates continuous and/or discrete (i.e., integer) decision variables, optimization problems are classified as *linear programming* (LP), *nonlinear programming* (NLP), *integer programming* (IP), and *mixed-integer (nonlinear) programming* (MI(N)LP) problems.

Algorithms for the solution of optimization problems are usually classified as *deterministic* (or *rigorous*) and *metaheuristic*. For an extensive introduction to optimization, the interested reader is referred to the following books: Nocedal and Wright (2000) give a broad overview of optimization theory and deterministic solution concepts. Edgar et al. (2001) cover the same contents with particular focus on chemical engineering problems. Spall (2003) and Brownlee (2011) give an exhaustive review of metaheuristic optimization algorithms. A comprehensive review of algorithms for the solution of mathematical optimization problems in the field of energy and process synthesis is given by Biegler and Grossmann (2004), and Grossmann and Biegler (2004).

In the following two sections, deterministic and metaheuristic optimization algorithms are briefly introduced. The discussion is based upon the listed references. It is assumed that the objective function is minimized for optimization.

2.2.1 Deterministic optimization algorithms

Deterministic optimization algorithms follow a predetermined search pattern for the solution of optimization problems. In 1947, George B. Dantzig developed the *simplex algorithm* for linear programs. For linear optimization problems, the optimal solution always lies on the boundary of the feasible region defined by the constraints. The simplex algorithm follows the edges of the polytope spanning the feasible region until it reaches the vertex with the optimal objective function value. Importantly, the identified solution is guaranteed to be the global optimal solution, because linear optimization problems are by definition convex, and in convex optimization problems, any local optimal solution is guaranteed to be globally optimal.

For nonlinear optimization problems, the optimal solution is generally not found on the boundary, but within the feasible region. Solution algorithms usually employ derivatives of the objective function to search for optimal solutions. There has been significant progress in the development of NLP algorithms (Biegler, 2010), e.g., SNOPT (Gill et al., 2005) and IPOPT (Wächter and Biegler, 2006). However, all NLP algorithms generally require good initial solutions to achieve convergence. Moreover, if the optimization problem is not convex, an identified optimal solution is not necessarily the global optimal solution, but the search can get stuck in local optimal solutions, and thus NLP algorithms cannot guarantee global optimality.

For integer programs, Land and Doig (1960) proposed the *branch-and-bound* method that systematically enumerates a subset of candidate solutions while discarding a large number of unqualified solutions. To do so, the branch-and-bound method recursively *branches* (i.e., partitions) the overall problem into a set of subproblems. Starting with the original problem, the algorithm solves an LP relaxation of each integer problem. For each relaxed variable of the generated solutions that takes a noninteger value, the problem is branched into a set of subproblems: Consider the relaxed variable y that takes a continuous value c, then we know that either $y \leq \lfloor c \rfloor$ (the smallest following integer value), or $y \geq \lceil c \rceil$ (the largest previous integer value); hence, two subproblems are generated, in each of which one of these constraints is added. For minimization, the objective function values of the subproblems represent *lower bounds* of the optimal solution's objective function value. With this information, the branch-and-bound algorithm can decide, which subproblems can be excluded from consideration or require further investigation, i.e., branching. This way, the problem can be solved to global optimality or until a termination criterion is met. As termination criterion, the user can specify the *maximal (relative) optimality gap*, which represents the tolerance between the so far best feasible solution (the *upper bound* for the optimal value) and the best known lower bound; a zero maximal optimality gap guarantees global optimality.

For mixed-integer programs, the branch-and-bound algorithm can be applied just as well by incorporating the continuous variables in the subproblems; for mixed-integer linear programs, it still guarantees global optimality. On the other hand, if discrete variables are added to nonlinear programming problems, again, conventional solution algorithms can guarantee global optimality only under simplifying assumptions, such as linearity of the discrete variables, and convexity of the nonlinear functions (Duran and Grossmann, 1986). To enable *global optimality* of general (MI)NLP problems, in recent years, large effort has been devoted to the development of *global optimization algorithms* (Horst and Tuy, 2010). Global optimization algorithms use branch-and-bound like techniques to decompose large and complex (MI)NLP problems into a set of smaller subproblems (Tawarmalani and Sahinidis, 2004, 2005; Misener and Floudas, 2012), which are then further analyzed to a) either show that no feasible or optimal solution exists, or b) identify the provably global optimal solutions to these subproblems, or c) further decompose the subproblem into more subproblems to be analyzed. In theory, this strategy will identify the global optimal solution of nonlinear problems (Falk and Soland, 1969; McCormick, 1976). However, for real-world problems, global optimization algorithms are mostly still not capable of identifying the global optimal solution within, for engineering practice, acceptable computation times (Rebennack et al., 2011). Moreover, they require exhaustive user inputs like feasible upper and lower bounds on all decision variables. Thus, given both the robustness issues of NLP algorithms and the high computational expense for global optimization, it is generally accepted that, if possible, nonlinearities should be avoided in the formulation of the mathematical programming problem to enable robust linear optimization (Biegler et al., 1997; Klatt and Marquardt, 2009; Grossmann, 2012).

2.2.2 Metaheuristic optimization algorithms

In this thesis, metaheuristics are considered that use randomized search patterns to thoroughly explore the search space. To progress the search, metaheuristic methods iteratively try to improve a candidate solution, and thus do not need to analyze the mathematical structure of the optimization problem. Therefore, they can be applied to problems, of which 1) the mathematical formulation is not known (black-box models), 2) the calculation of derivatives is excessively complex, or 3) deterministic methods fail. The major advantage of metaheuristic search algorithms is that, if properly designed, they do not get stuck in local optimal solutions. The major disadvantage of metaheuristics is that they can have very slow convergence and usually need to be tuned to a particular problem. For optimization-based synthesis, mostly evolutionary algorithms and simulated annealing methods have been used (Baños et al., 2011); the latter, however, primarily during the 1980s and 90s of the last century (Kirkpatrick et al., 1983; Cerný, 1985). In the following, evolutionary algorithms are discussed in more detail.

Evolutionary algorithms

Evolutionary algorithms (EAs) form a subset of metaheuristic optimization methods inspired by biological evolution (Eiben and Smith, 2003). EAs are derivative-free methods for numerical optimization of (non-)linear, (non-)convex mixed-integer optimization problems. EAs are regarded as global optimization methods although convergence towards the global optimal solution can generally not be guaranteed for a limited number of objective function evaluations. However, EAs typically perform well on multi-modal and nonlinear functions. Moreover, a well-designed EA will usually not get stuck in local optimal solutions, but will find good and even near-optimal solutions. Prominent implementations in the field of energy and process synthesis are evolution strategies (ESs) and genetic algorithms (GAs) (Pezzini et al., 2011).

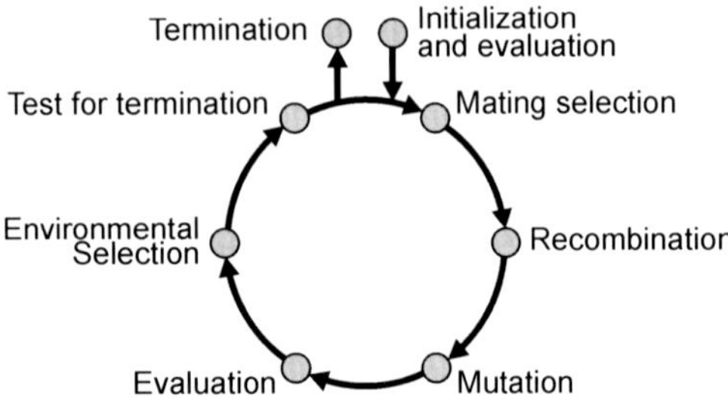

Figure 2.4: Flow diagram of an evolutionary algorithm (adapted from (Eiben and Smith, 2003)).

The basic run of an EA is shown in Fig. 2.4. The algorithm starts with a one-time *initialization* and *evaluation*: A set (*population*) of candidate solutions (*individuals*) is randomly created and evaluated using an objective function (*fitness function*). Afterwards, the algorithm starts an iteration loop: Based on their fitness values, *parent individuals* are selected to create new individuals by *mating selection*. The creation of new individuals is performed by *recombination* and *mutation*: The recombination operator randomly takes and reassembles parts from the selected parents to create new offspring; the mutation operator randomly alters single parent individuals to create new individuals. The newly born offspring is *evaluated*, and finally, *environmental selection* kills those individuals from the population that will least likely evolve into the optimal solution. Selection, recombination, and mutation are called *genetic operators*. It should be noted that recombination is not essential for the successful application of an EA to an optimization problem, but can be used to enhance the randomized search. In contrast, mutation is indispensable for every EA. An iteration of the described loop is called a *generation*. An EA continues this evolution process until a termination criterion is met, e.g., a maximum number of generations, or a maximum number of generations without improvement of the objective function. It should be pointed out that, during this evolution process, EAs will generate a range of solutions rather than only a single optimal solution.

Different EAs employ different selection strategies. A well-known selection strategy is the $(\mu + \lambda)$-selection (Beyer and Schwefel, 2002): In each generation, λ offspring individuals are generated from a set of μ parent individuals. Selection compares the fitness values of all $\mu + \lambda$ individuals and discards the λ worst. The "+"-selection guarantees survival of the best individuals found so far. The algorithm is commonly configured such that $1 \leq \mu \leq \lambda$ (Brownlee, 2011). The ratio of μ to λ represents the selection pressure, or greediness, applied by the algorithm.

2.3 Optimization-based synthesis methods

In this section, a review is provided of optimization-based synthesis methods including a discussion of drawbacks and shortcomings of the available methods.

2.3.1 Superstructure-based synthesis methods

Based on the early optimization approaches for energy and process systems synthesis (e.g., by Nishio and Johnson (1979) and Nishida et al. (1981)), Westerberg (1991) proposed the general framework for *superstructure-based* process synthesis, which has been exhaustively applied to many different process synthesis problems. More recently, it has been generalized to the synthesis of energy systems as reviewed by (Liu et al., 2011).

The basic concept of superstructure-based synthesis is to identify the optimal solution among all alternatives embedded in one integrated *superstructure*, which includes all possible interconnections among the considered technical components. According to the definition by Westerberg (1991), the problem is decomposed into three tasks that need to be addressed sequentially:

1. Definition of a superstructure, which encodes the synthesis alternatives,
2. mathematical modeling of the superstructure, and
3. solution of the optimization problem employing mathematical programming techniques.

As discussed in section 2.1.2, the synthesis of distributed energy supply systems has to be considered on three levels: the synthesis, the design, and the operation level. Accordingly, the general superstructure optimization problem for distributed energy supply systems is given by an MINLP:

$$\begin{aligned} \min_{x,y} \quad & f(x,y), \\ \text{s.t.} \quad & h(x,y) = 0, \\ & g(x,y) \leq 0, \\ & x \in \mathbb{R}^n, \\ & y \in \{0,1\}^m, \end{aligned} \tag{2.2}$$

where x is a vector of n continuous decision variables representing flow rates, equipment sizing, temperatures, pressures, etc.,

y is a vector of m binary decision variables denoting the (non-)existence and the on/off-status of a piece of equipment in the superstructure,

$f(x, y)$ is the objective function,

$h(x, y)$ are the equality constraints modeling the systems constraints, i.e., mass balances, energy balances, etc., and

$g(x, y)$ are the inequality constraints representing design specifications, feasibility constraints, logical constraints, and other constraints.

In the following, a brief review is given of mathematical programming formulations for the superstructure-based optimization of (distributed) energy supply systems: Nishio and Johnson (1979) proposed the first mathematical programming approach for the optimal synthesis of steam power plants employing an LP model. However, this approach uses heuristics for some decisions on the system structure. Moreover, only strictly linear investment cost relations are assumed, thus neglecting the economy of scale of equipment cost. To avoid these shortcomings, Papoulias and Grossmann (1983) proposed an MILP approach for the optimal synthesis of energy supply systems considering piecewise linearized investment cost curves. But this approach was limited to model stationary system behavior, i.e., constant demands. Iyer and Grossmann (1997) proposed an MILP approach for the multi-period operation optimization of energy supply systems. However, both approaches model equipment performance to be depending only on equipment size, and thus constant efficiencies are assumed over the whole range of operation. This assumption significantly limits the potential solution space of the optimization problem and generally leads to suboptimal solutions. Prokopakis and Maroulis (1996) proposed an MINLP approach for the operation optimization of energy supply systems taking into account size- and load-dependent equipment performance. Papalexandri et al. (1998) and Bruno et al. (1998) generalized the MINLP formulation to the optimal synthesis of energy supply systems. More recently, very detailed equipment models have been proposed for the MINLP operation optimization (Varbanov et al., 2004; Velasco-Garcia et al., 2011) and optimal synthesis (Varbanov et al., 2005; Chen and Lin, 2011) of energy supply systems. While the MINLP approaches are capable of simultaneously taking into account all characteristics inherent to distributed energy supply systems, the solution of the corresponding optimization problems is anything but a routine task. Thus, for the synthesis of practically relevant problems, today, mostly still MILP approaches are used. However, these approaches do not simultaneously optimize equipment configuration, sizing, and operation: Either, constant efficiencies are assumed over the whole range of operation (Söderman and Pettersson, 2006; Piacentino and Cardona, 2008; Kang et al., 2011; Chen et al., 2011; Keirstead et al., 2012); or, equipment performance is modeled to be depending only on the operation loads, and thus no continuous equipment sizing is performed (Hui and Natori, 1996; Casisi et al., 2009; Buoro et al., 2011). Yokoyama et al. (2002) proposed an MILP formulation capable of simultaneously optimizing equipment configuration, sizing, and operation taking into account time-dependent demands, continuous equipment sizing, and part-load operating performance.

Besides these deterministic optimization approaches, many metaheuristic optimization approaches have been proposed for the superstructure-based energy systems synthesis. Among them are methods based on evolutionary algorithms (Uhlenbruck and Lucas, 2004; Koch et al., 2007; Dimopoulos and Frangopoulos, 2008), particle swarm algorithms (Eberhart and Shi, 2001; Soares et al., 2012), and leapfrogging algorithms (Rhinehart et al., 2012).

Drawbacks of superstructure-based synthesis methods

In the past, strenuous efforts have been made to provide user-friendly frameworks to facilitate the use of superstructure-based optimization, such as PROSYN (Kravanja and Grossmann, 1993) and Jacaranda (Fraga, 1998; Fraga et al., 2000). Current optimization software for optimal systems synthesis have been reviewed by Caballero et al. (2007), Kravanja (2010), and Lam et al. (2011) showing that optimization frameworks are available, which allow for graphical modeling of superstructures. These tools employ modular, component-based modeling approaches, thus facilitating the use of optimization for systems synthesis. However, the major issue inherent to the superstructure-based synthesis remains: the manual modeling of an adequate superstructure for a given synthesis problem (Andrecovich and Westerberg, 1985). The designer needs to decide *a priori*, which alternatives should be included in the superstructure, thereby running the risk to exclude the optimum from consideration. Principally, this risk can be circumvented by using excessively large superstructures; but then again, the complexity of optimal synthesis problems increases exponentially with the number of equipment considered in the superstructure (Kallrath, 2000), and thus unnecessarily large superstructures lead to prohibitive computational effort. Westerberg (2004) concludes that the only way to search large design spaces of complex synthesis problems is to analyze and exploit the special structure and properties of the synthesis problem at hand. For this purpose, efficient superstructure models have been proposed, e.g., for the synthesis of separation systems (Floudas, 1987; Kovács et al., 2000), and the synthesis of steam turbine systems (Mavromatis and Kokossis, 1998). However, these models are specifically tailored to particular synthesis problems, and thus very limited in their scope of application.

To circumvent the problems related to the manual definition of superstructures, in the past, two alternative approaches have been developed:

- The algorithmic generation of superstructures, and
- the use of superstructure-free synthesis methods.

Algorithms for superstructure generation automatically define a superstructure for a given synthesis problem. Superstructure-free synthesis approaches are closely related to the concept of evolutionary programming (Hillermeier et al., 2000): Superstructure-free synthesis methods start from some initial structure and employ mutation operators that gradually refine the structure by either pruning, adding, or reconnecting parts of the flowsheet. In the following, both classes of methodologies are reviewed in more detail.

2.3.2 Superstructure generation methods

For process synthesis, two algorithmic methods have been proposed for superstructure generation: Friedler et al. (1992) introduced the *P-graph* based *PNS framework* (process network synthesis), which employs a generic graph representation of process systems to automatically generate and optimize superstructures for process synthesis. More recently, d'Anterroches and Gani (2005) presented the *group contribution based process flowsheet synthesis* methodology for the generation of process superstructures. As its name implies, the group contribution based methodology draws an analogy to group contribution methods for pure component property prediction. The group contribution based methodology is specifically tailored to process synthesis problems, and thus it cannot be easily transferred to energy systems synthesis problems. In contrast, the P-graph based PNS framework can be easily modified for its application to energy systems synthesis as shown by Varbanov et al. (2011). In the literature, no superstructure generation algorithm has been proposed that is specifically tailored to the synthesis of energy supply systems. Thus, in this thesis, the PNS framework is employed as basis for automated superstructure generation for the synthesis of distributed energy supply systems. In the following, the PNS framework is presented in detail. A thorough introduction to the PNS framework is given by Peters et al. (2003) and Klemes et al. (2011).

P-graph based PNS framework

A *P-graph* is a graph-theoretic process model for unambiguous representation of synthesis alternatives, i.e., solution structures. A process synthesis problem is defined by the triplet (P, R, O) where P is the set of required products, R is the set of available raw materials, and O is the set of available operating units. P and R are subsets of the more general set M representing all materials including intermediate materials that might be produced and used by the operating units. An operating unit O is the set of pairs of two subsets of M representing the input and output materials of the operating unit. For efficient and unambiguous definition of all feasible solution structures, a set of axioms has been proposed that defines necessary structural properties of feasible process systems (Friedler et al., 1993):

(S1) Every final product is represented in the graph.

(S2) A vertex of the M-type has no input if and only if it represents a raw material.

(S3) Every vertex of the O-type represents an operating unit defined in the synthesis problem.

(S4) Every vertex of the O-type has at least one path leading to a vertex of the M-type representing a final product.

(S5) If a vertex of the M-type belongs to the graph, it must be an input to or an output from at least one vertex of the O-type in the graph.

Axiom (S1) requires that each product is produced by at least one of the operating units of the system. Axiom (S2) implies that a material is not produced by any operating unit of the system if this material is available as raw material. Axiom (S3) implies that only those operating units are taken into account that can be reasonably connected to materials defined by the operating units' inputs and outputs. Axiom (S4) requires that any operating unit of the system has at least one path leading to a material representing a product. And finally, axiom (S5) implies that each material of the system is an input to or an output from at least one operating unit of the system. A solution structure complying with these axioms is said to be *combinatorially feasible*.

For illustration, consider the following synthesis problem (Fig. 2.5). The material set is $M = \{A, B, C, D, E, F, G\}$ defined by the operating units $O = \{(\{B, C\}, \{A, F\}), (\{A, E\}, \{C\}), (\{D\}, \{C\}), (\{C, E\}, \{G\})\}$ with $R = \{E\}$ and $P = \{F\}$. The P-graph that incorporates the given materials and operating units (Fig. 2.5 b) is not combinatorial feasible. In fact, according to the P-graph axiom (S2), in the simple example, material D is no raw material, and thus operating unit O_2 is discarded. Moreover, operating unit O_4 is omitted because it has no connections eventually leading to any of the products, thus violating axiom (S4). The combinatorial feasible P-graph is shown in Fig. 2.5 c).

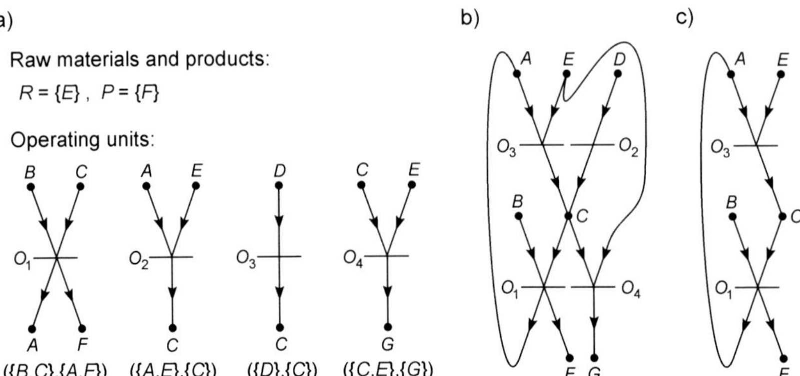

Figure 2.5: Simple example of a synthesis problem represented as P-graph. a) Problem definition. b) Complete P-graph. c) P-graph incorporating only combinatorially feasible solution structures.

The well-established mathematical foundation of graph theory enables to efficiently screen P-graphs according to the axioms to decide which encoded structures are combinatorially feasible, and which structures must be discarded from the P-graph. The P-graph based *PNS* framework provides algorithmic methods for both generation (Friedler et al., 1993) and optimization (Friedler et al., 1996) of mathematical superstructure models. The algorithm for maximal structure generation (MSG) generates the so-called *maximal structure*, a superstructure proven to contain all combinatorially feasible synthesis alternatives to a given process synthesis problem. In particular, the maximal structure always contains exactly one unit of each *plausible* equipment type, i.e., each equipment that can be reasonably connected within the combinatorially feasible synthesis alternatives.

The P-graph approach has been exhaustively applied to classic process synthesis problems (Fan et al., 2012; Heckl et al., 2010; Lin et al., 2009; Fan et al., 2009, 2008). Recently, it has also been applied to the synthesis of regional supply chains (Lam et al., 2010) and utility systems (Halasz et al., 2010; Varbanov et al., 2011). For energy systems synthesis, analogies are drawn between materials and energy flows, and operating units and energy conversion technologies. However, the PNS framework has originally been designed for synthesizing chemical plants, and thus neglects major characteristics of distributed energy supply systems that have a strong impact on these systems' performance as demonstrated in the introductory example and discussed by Velasco-Garcia et al. (2011) and Zhou et al. (2013):

First of all, the original PNS framework does not consider multiple redundant units as usually necessary for DESS (cf. section 2.1.2). To incorporate redundant units in a P-graph superstructure, manual manipulations are necessary. Recently, Bertok et al. (2013) proposed an extension of the PNS framework to consider *redundant structures* for better reliability of supply chains. However, this approach only accounts for alternative pathways in supply chains providing the same materials, but not for redundant technical units within one pathway. Moreover, the approach still requires user inputs for the definition of possible redundant structures.

In addition, the PNS framework does not consider time-varying constraints (e.g., demand profiles, and ambient temperature curves). Moreover, it assumes constant equipment efficiencies, and hence neglects part-load dependent equipment performance. In summary, in its current state, the P-graph based PNS framework is not suitable for the synthesis of distributed energy supply systems.

2.3.3 Superstructure-free synthesis methods

To avoid the *a priori* definition of a superstructure, superstructure-free synthesis approaches merge the paradigms for systematic superstructure generation and optimization.

For heat exchanger network synthesis, Androulakis and Venkatasubramanian (1991), and Lewin et al. (1998) proposed bi-level approaches that employ genetic algorithms for the configuration evolution and deterministic methods for the heat load dispatch. Luo et al. (2009) introduced a similar procedure combining a genetic algorithm for structure optimization, and a simulated annealing algorithm for parameter optimization. Fraga (2009) proposed a string rewriting grammar (Book, 1987) for the network evolution. The concept of string rewriting is closely related to the more general concept of graph grammars, which is employed in this thesis (cf. section 5.3). All approaches are specifically tailored to the synthesis of heat exchanger networks, and thus cannot be applied to DESS synthesis problems.

Angelov et al. (2003) and Wright et al. (2008a,b) proposed a methodology for the superstructure-free synthesis of heating, ventilating and air conditioning (HVAC) systems. The proposed method uses genetic operators tailored to the synthesis of HVAC systems. For mutation, two operators have been developed: One operator swaps interconnections of two randomly selected components; another operator exchanges two randomly selected components. For recombination, again two operators have been designed: One operator creates offspring individuals retaining most of the structural properties of one parent, and technical specifications from the other parent; the other operator gives birth to offspring individuals retaining properties of both parents in equal measure. The approach has been successfully applied to the optimization of HVAC systems. It employs a generic component-based modeling framework incorporating multiple different technologies. However, this approach is not capable of evaluating replacement investments for the installation of new components.

Emmerich et al. (2001) proposed a more flexible approach for the superstructure-free synthesis of energy and process systems. The approach uses a mutation operator that applies a multitude of technology-specific replacement rules. The mutation operator is based on so-called *minimal moves*. A minimal move recognizes existing structural patterns of a flowsheet and randomly replaces them by other, similar patterns; e.g., "*Insert a pump parallel to an existing pump*", or "*Swap a stabilizer column with a cooler and flash combination*". To avoid the introduction of any bias in the mutation process, an inverse counterpart is provided for each minimal move. The proposed method has been applied successfully to the optimization of chemical processes (Emmerich et al., 2001; Urselmann et al., 2007; Sand et al., 2008) and thermal power plants (Emmerich, 2002). The major drawback of the minimal moves approach is that the designer needs to define a multitude of technology-specific replacement rules. This is anything but a trivial task as an illustrative example shows: Imagine the minimal move "*Insert a chiller unit in parallel to an existing chiller unit*". This rule does not yet describe the interconnection of the two units with respect to their cold and cooling water cycles. Thus, for this specific minimal move, a set of three different replacement rules needs to be implemented (plus all of their inverse counterparts): one rule for parallel connection of the cold water cycles, one rule for parallel connection of the cooling water cycles, and one rule for parallel connection of both the cold and the cooling water cycles.

The listed superstructure-free methods involve several drawbacks that either impede their use for the synthesis of distributed energy supply systems, or due to which these methods are not superior to the traditional superstructure-based synthesis approach: The methods for heat exchanger network synthesis are too specific to be easily transferred to the synthesis of DESS. The methodology for HVAC systems synthesis neglects the possibility of equipment substitution. And the minimal moves concept requires manual definition of many technology-specific replacement rules, which is at least equally, if not more, difficult as the manual superstructure definition.

2.4 Multi-objective optimal synthesis

For real-world synthesis problems, it is often not sufficient to optimize a single objective function. Instead, multiple criteria (economic, environmental, social, etc.) need to be considered. Østergaard (2009) reviews several criteria for the optimal synthesis of energy systems. He shows that there is no generally accepted common optimization criterion. In fact, in case of single-objective optimization, the choice of the optimization criterion has a strong impact on the outcome of the optimization, and thus, by choosing a single objective function, the decision maker introduces possibly undesired preferences into the optimization process. Moreover, it is generally not possible to identify a single optimal solution that simultaneously optimizes all objectives. To address these issues, *multi-criteria decision analysis* and *multi-objective optimization* (MOO) approaches have been developed. A comprehensive introduction to these methods is given in the books by Miettinen (1999) and Ehrgott et al. (2010), upon which the following section is based.

In contrast to single-objective optimization techniques, MOO techniques perform vector optimization. The goal is to identify the solution vector representing the best possible trade-offs in the objective function space. Accordingly, a multi-objective minimization problem is formulated as follows:

$$\begin{aligned} \min_{x} \quad & (f_1(x), f_2(x), \dots f_k(x))^{\intercal} \\ \text{s.t.} \quad & x \in S, \end{aligned} \tag{2.3}$$

with k objective functions $f : \mathbb{R}^n \to \mathbb{R}$. The vector of objective functions is denoted by $f(x) = (f_1(x), f_2(x), \dots, f_k(x))^T$. The decision variable vector $x = (x_1, x_2, \dots, x_n)^T$ belongs to the (nonempty) feasible region S. This notation is used throughout the remainder of this section (including the assumption that the objective functions have to be minimized).

The result of an MOO problem is a set of so-called *Pareto-optimal* solutions, all of which define the *Pareto-set*. The set of the generated Pareto-optimal solutions is called the *Pareto-front*. For the MOO problem (2.3), a feasible solution $\hat{x}$ is Pareto-optimal, if there is no other feasible solution x subject to

$$f_i(x) \leq f_i(\hat{x}) \quad \forall\, i = 1, \dots, k \tag{2.4}$$

with at least one strict inequality. Simply put, a solution is Pareto-optimal, if no other solution can be found that improves at least one of the objective functions without deteriorating the performance in any other objective function. Superior solutions *dominate* inferior solutions. According to this definition, the Pareto-set represents all reasonable actions a rational decision maker can take. Thus, MOO methods integrate optimization and decision support. Therefore, MOO methods aim at 1) the approximation of the Pareto set, and 2) finding solutions that are widely spread throughout the objective function space.

Traditionally, three classes of methods are distinguished for solving MOO problems depending on when the decision maker is involved to express decision preferences (Hwang and Masud, 1979):

A priori methods. *A priori* methods require information on the decision maker's preferences before the solution process starts. These methods essentially transform the MOO problem into a single-objective optimization problem. A well-known *a priori* method is the *weighting method*.

Interactive methods. Interactive methods require information on the preferences of the decision maker throughout the solution process to direct the search and refine existing solutions.

A posteriori methods. *A posteriori* methods generate Pareto-optimal solutions to approximate the Pareto set. The decision maker analyzes the generated solutions to choose the one with the preferred trade-offs. Commonly used *a posteriori* methods are the *weighting method*, the *ε-constraint method* and *evolutionary multi-objective optimization algorithms* (EMOAs). It should be noted that the analysis of the generated solution set can be a complex task itself, and therefore specific analysis methods have been developed, which are described in detail by Belton and Stewart (2002) and Figueira et al. (2005).

For the synthesis of distributed energy supply systems, almost exclusively *a posteriori* methods are used. Reviews of multi-objective synthesis methods are given by Toffolo and Lazzaretto (2002), Pohekar and Ramachandran (2004), Wang et al. (2009), and Alarcon-Rodriguez et al. (2010). Among others, Guillen-Gosalbez et al. (2008), Mavrotas et al. (2008) and Liu et al. (2010a,b) employ the ε-constraint method. Weber et al. (2006), Bouvy and Lucas (2007), Maréchal et al. (2008), Kavvadias and Maroulis (2010), and Kusiak et al. (2011a) use EMOAs. Fazlollahi et al. (2012) applies both an EMOA and an ε-constraint-based method: For the considered case study, the ε-constraint method is computationally less involved than the EMOA; however, the EMOA identifies a wider range of (near-)Pareto-optimal solutions. Besides EMOAs, other metaheuristic algorithms have recently been designed for the solution of multi-objective energy systems synthesis problems, e.g., particle-swarm (Kusiak et al., 2011b) and leapfrogging (Niknam et al., 2011) algorithms.

In the following, the three most prominent representatives of MOO algorithms are described in detail: the weighting method, the ε-constraint method, and evolutionary multi-objective algorithms (EMOAs).

2.4.1 Weighting method

The weighting method can be used as both an *a priori* and an *a posteriori* method. Multiple objective functions are transformed into a single objective function by weighting each objective function and minimizing the weighted sum of the objective functions:

$$\begin{aligned} \min_{x} \quad & \sum_{i=1}^{k} w_i \, f_i(x) \\ \text{s.t.} \quad & x \in S, \end{aligned} \tag{2.5}$$

where the weighting coefficients w_i are typically real positive numbers ($w_i \geq 0$ for all $i = 1, \ldots, k$), usually normalized such that $\sum_{i=1}^{k} w_i = 1$.

In case of the *a priori* weighting method, the decision maker specifies the weighting coefficients before the computation starts. This method is commonly used, e.g., the net present value weights capital cost and running expenses in economic optimization. For its use as *a posteriori* method, the values of the weighting coefficients are systematically changed to approximate the Pareto set. It should be noted that solutions lying on an edge connecting two Pareto-optimal solutions are generally not Pareto-optimal themselves. Moreover, for nonconvex MOO problems, it is not guaranteed that the weighting method is even capable of identifying all Pareto-optimal solutions.

2.4.2 ε-constraint method

Haimes et al. (1971) proposed the ε-constraint method, which converts the multi-objective optimization problem into a set of single-objective optimization problems. For this purpose, the ε-constraint method performs optimization with respect to one of the multiple objective functions while upper or lower bounds are set for all other objective functions:

$$\begin{aligned} \min_{x} \quad & f_l(x) && \forall\, l = 1, \ldots, k, \\ \text{s.t.} \quad & f_j(x) \leq \varepsilon_j && \forall\, j = 1, \ldots, k, \quad j \neq l, \\ & x \in S. \end{aligned} \tag{2.6}$$

For multi-objective optimization, the objective functions to be minimized (or maximized) and the bounds of the constraints are systematically changed to approximate the Pareto set.

It should be noted that, in contrast to the weighting method, theoretically, the ε-constraint method can identify every Pareto-optimal solution of any MOO problem. However, because of the increasing number of constraints, the ε-constraint method is computationally more involved than the weighting method. Another difference worth noting is that while for the weighting method a lot of runs might be spent to identify different solutions, the ε-constraint method generates in almost every run different Pareto-optimal solutions, and thus it can be used to control the number of generated Pareto-optimal solutions. Therefore, the ε-constraint method enables to generate a richer approximation of the Pareto set.

2.4.3 Evolutionary multi-objective algorithms (EMOAs)

Zitzler and Thiele (1998) and Coello et al. (2007) review evolutionary multi-objective algorithms (EMOAs). EMOAs are particularly suited for MOO if they exploit their population-based solution procedure. Coello et al. (2007) classify these EMOAs as *criterion selection techniques* and *aggregation selection techniques*:

Schaffer (1985) proposed a criterion selection technique that is generally considered the first implementation of an EMOA. The basic concept of the *Vector Evaluated Genetic Algorithm* (VEGA) is to generate k sub-populations for each of the k objectives to be optimized. Each sub-population is optimized with respect to one of the k objective functions. For mating selection, the sub-populations are shuffled and re-partitioned. Finally, recombination, mutation, and environmental selection are applied to the new population as for a standard single-objective EA.

Hajela and Lin (1992) and Ishibuchi and Murata (1998) proposed *aggregation selection* EMOAs that employ the weighting method. However, rather than using constant weights for the objective functions, the weights are varied during optimization; from generation to generation and/or from function evaluation to function evaluation. Depending on the specific implementation, the weights are assigned randomly, or according to some heuristic strategy.

More recently, Beume et al. (2007) proposed the *S-metric selection* EMOA (SMS-EMOA). The SMS-EMOA enables computationally efficient and thorough exploration of multi-objective solution spaces. It applies a $(\mu + 1)$-selection strategy; i.e., the population constantly consists of μ individuals, and in each generation, only one offspring individual is created. To decide which individuals will survive to the next generation, selection calculates the *S-metric* for each solution. The S-metric is an integral value of the hypervolume spanned from the solution to the so-called *reference point*. The reference point is an approximation of the so-called *nadir point* (Steuer, 1986), which defines the worst *possible* solution, i.e., the solution with the worst possible objective function values in each objective. The reference point serves as upper bound for the population's objective function values and for orientation to assess the distribution of the population. For two objectives, the spanned hypervolume is an area in the objective function value space, cf. Fig. 2.6. The S-metric assumes that solutions with larger hypervolumes will more likely evolve into Pareto-optimal solutions than solutions with smaller hypervolumes, and therefore the former are kept in the population while the latter are discarded. Selection also rewards diversity, i.e., a representative distribution of solutions along the Pareto-set. Thus, depending on the solutions' positions in the objective function value space, the algorithm might discard some dominant individuals while other dominated individuals might be kept for the benefit of preserving a distributed solution set.

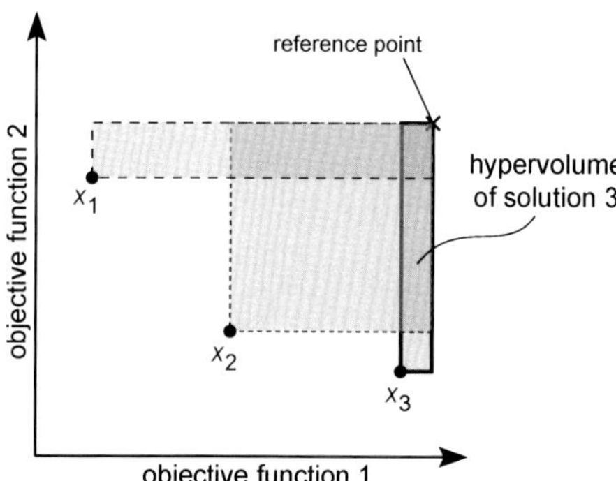

Figure 2.6: Objective function space of two objective functions. Three dominant solutions are plotted with their hypervolumes. Solution 3 spans the smallest hypervolume to the reference point.

2.5 Contribution of this thesis

The review given in this chapter reveals shortcomings of the available optimization-based methods for the synthesis of distributed energy supply systems. Therefore, in this thesis, two novel optimization-based synthesis methodologies are proposed that meet all of the discussed requirements at the same time. The scope of this thesis is the synthesis of distributed energy supply systems on the conceptual level.

Generic automated synthesis methodologies. In this thesis, two novel methodologies are presented for the automated optimization-based synthesis of distributed energy supply systems: a superstructure-based and a superstructure-free synthesis methodology.

In chapter 4, the superstructure-based synthesis methodology is proposed, which is based on a framework for automated superstructure generation and optimization. To identify the optimal solution, the framework employs a successive superstructure expansion and optimization strategy that continuously increases the number of units included in the superstructure until the optimal solution is identified.

In chapter 5, the superstructure-free optimization methodology is proposed. In search for the optimal solution, this methodology simultaneously generates and optimizes candidate solutions. The approach is based on a knowledge-integrated evolutionary algorithm that applies a handful of generic replacement rules for the evolution of solution structures. For this purpose, all relevant technologies are classified into a hierarchically-structured graph that allows for an efficient and meaningful definition of all reasonable connections, thus minimizing the number of replacement rules. Moreover, this approach automatically defines all replacement rules.

The proposed methodologies avoid both the *a priori* definition of a superstructure and the manual definition of many technology-specific replacement rules. Moreover, both methods consider multiple redundant units as generally necessary for the optimal synthesis of distributed energy supply systems. Furthermore, both methods employ a generic component-based modeling to enable the use of different modeling frameworks.

Considering the major characteristics of distributed energy supply systems. In this thesis, for system modeling, every single energy conversion unit is explicitly considered[2]. This modeling approach enables detailed analysis of the systems synthesis

[2]This is in contrast to urban- (e.g., Keirstead and Shah (2011)) and country-scale models (e.g., Fishbone and Abilock (1981)), for which the set of energy conversion units is mostly represented by aggregated models reflecting only the total capacities and average efficiencies.

including the single units' sizing and operation. A robust MILP formulation is used to simultaneously optimize the structure, sizing, and operation of distributed energy supply systems (chapter 3). The MILP formulation is similar to the one proposed by Yokoyama et al. (2002) and accounts for time-varying load profiles, continuous equipment sizing, economy of scale of equipment investment, and part-load equipment performance. At the same time, it includes already existing equipment in the problem formulation, thus allowing to perform both grassroots and retrofit synthesis. Moreover, as an extension of this thesis, topographic constraints can be defined to take into account constructional limitations on the considered site.

Generation of near-optimal solution alternatives. Optimization models commonly suffer from the following two shortcomings: First, a mathematical model is never a perfect representation of the real world, and the decision maker is often not aware *a priori* of all constraints relevant for the synthesis problem at hand. Thus, the optimal solution is usually only an approximation of the optimal real-world solution. Second, in rapidly changing economic and technological environments (e.g., varying energy tariffs, changing energy demands, etc.), the constraints will change in the future. However, the optimal solution generally only reflects the status quo of the current situation. Due to these shortcomings, a single optimal solution does not suffice in practice. Instead, a deeper understanding of the synthesis problem at hand is required to reflect the real-world situation. For this purpose, the proposed synthesis frameworks implement strategies to generate a set of near-optimal solution alternatives rather than only a single solution. The near-optimal solution alternatives can be evaluated in more detail *a posteriori*, thus providing deeper understanding of the intrinsic nature of the synthesis problem and opening up a wider space for rational synthesis decisions. In summary, this approach supports the decision maker to account for aspects that have not been explicitly considered during optimization.

Multi-objective decision support. Single-objective optimization does not account for multiple (contradicting) criteria as necessary for real-world problems: For instance, economic optimization with respect to a single aggregated objective function, e.g., the net present value, requires to assign fixed weights to the investments and running expenses. Thus, by choosing the weights, the decision maker introduces possibly undesired *a priori* preferences into the optimization process. To rigorously consider multiple criteria, the proposed synthesis methodologies implement strategies for multi-objective optimization to generate a set of Pareto-optimal solutions.

Real-world synthesis. In chapter 6, both the superstructure-based and the superstructure-free synthesis frameworks are applied to a real-world synthesis problem. The synthesis problem is addressed by a three-step synthesis procedure employing the above-listed synthesis approaches: First, single-objective synthesis is performed maximizing the net present value. Secondly, a set of promising near-optimal solution alternatives is generated. Finally, multi-objective synthesis is performed considering two contradicting objectives: minimization of total investment cost and minimization of cumulative energy demand (CED). The features of both synthesis frameworks are discussed with regard to practicality, i.e., solution quality and computational performance, and a comparative evaluation is provided. It is shown that the presented three-step synthesis procedure based on the novel synthesis methodologies proposed in this thesis enables to identify both common features and differences among the generated solution alternatives, thus providing valuable insight into the real-world synthesis problem, thereby supporting the decision maker to reach rational and far-sighted synthesis decisions.

Comparative evaluation of deterministic and metaheuristic optimization. Most papers in literature either use deterministic or metaheuristic optimization, and therefore mainly point out the advantages of either of them. For this reason, fair comparisons are hardly available. In this thesis, a comparative evaluation is provided of both the deterministic superstructure-based and the metaheuristic superstructure-free synthesis methodologies.

Chapter 3

An MILP framework for modeling distributed energy supply systems

In this thesis, a multi-period modeling approach is employed for the synthesis of distributed energy supply systems (DESS). Quasi-stationary system behavior is assumed (a common assumption in literature (Pillai et al., 2011)); i.e., the system is assumed to reach steady-state conditions instantly based on the current constraints. Therefore, a time trajectory is modeled as a sequence of quasi-stationary operating points. The modeling approach limits the user-inputs to easily obtainable data, such as performance curves from manufacturer data sheets. This approach enables easy model parameterization, while, at the same time, yielding sufficiently accurate results for the analysis of energy systems and the development of improvement measures (Voll et al., 2010). In the next two sections, the equipment models and a mathematical programming formulation are presented for DESS synthesis. For robust optimization, i.e. reliable convergence, it is generally accepted that, if possible, nonlinearities should be avoided in the formulation of the mathematical programming problem (Biegler et al., 1997; Klatt and Marquardt, 2009). Thus, based on the work by Yokoyama et al. (2002), an MILP formulation is derived that simultaneously optimizes equipment configuration, sizing, and operation accounting for time-dependent demands, continuous equipment sizing, and part-load operating performance. In section 3.3, the proposed MILP formulation is evaluated with regard to its approximation of the original MINLP problem and required computational effort to solve a synthesis test problem. The chapter is concluded with a short summary. Major parts of this chapter are based on earlier publication by Voll et al. (2013).

3.1 Equipment models

The equipment models used in this work comprise investment cost functions, nominal efficiencies, and part-load performance curves. The component set encompasses boilers, combined heat and power (CHP) engines, turbo-driven compression chillers, and absorption chillers. If available, the required economic and technical parameters are taken from the German market (Gebhardt et al., 2002; Scheunemann and Becker, 2004), or else they were provided by industry partners. For simplicity, cooling towers are assumed to be part of the chiller units.

In Table 3.1 the capacity and price ranges of the considered technologies as well as their maintenance cost, and nominal overall efficiencies and coefficients of performance (COPs) are listed.

Table 3.1: Considered energy conversion technologies including their power and cost ranges, and nominal efficiencies η_{N} for boilers and CHP engines, and COPs for chillers (AC: absorption chiller, TC: turbo-driven compression chiller).

technology	thermal power range / MW	investment cost / k€	maintenance cost / % investment cost	η_{N}, $\mathrm{COP}_{\mathrm{N}}$ / -
Boiler	0.1 - 14.0	34 - 380	1.5	0.9
CHP engine	0.5 - 3.2	230 - 850	10.0	0.87
AC	0.1 - 6.5	75 - 520	4.0	0.67
TC	0.4 - 10.0	89 - 1570	1.0	5.54

Investment costs are calculated by capacity power laws (Smith, 2005):

$$I_{\mathrm{N}} = I_{\mathrm{B}} \cdot \left(\frac{\dot{Q}_{\mathrm{N}}}{\dot{Q}_{\mathrm{B}}} \right)^{M} , \tag{3.1}$$

in which I_{N} represents the cost for equipment with capacity $\dot{Q}_{\mathrm{N}}$, I_{B} represents the reference cost for equipment with capacity $\dot{Q}_{\mathrm{B}}$, and M is a constant parameter depending on the equipment type.

According to the gathered technical data, the nominal overall efficiencies are assumed to be independent of equipment size for each technology (Table 3.1). In contrast, the ratio of the CHP engine's nominal electric and thermal efficiencies changes with equipment size and is reflected by the model (cf. appendix A.1).

On the conceptual design level, only standardized equipment is considered. Thus, for the underlying performance models, it is justified to assume a characteristic behavior for all units of a certain technology. Therefore, all technologies' part-load efficiency curves are normalized to their nominal efficiencies. These characteristic performance curves are then assumed to be valid for a certain technology regardless of the specific units' capacities. The characteristic performance curves employed in this study are shown in Fig. 3.1. These include the relative boiler efficiency, η/η_{N}, the relative CHP engine's overall and electric efficiencies, η/η_{N} and $\eta_{\mathrm{el}}/\eta_{\mathrm{el,N}}$, and the chillers' relative coefficients of performance, $\mathrm{COP}/\mathrm{COP}_{\mathrm{N}}$. The minimum part-loads are assumed to be 50 % for CHP engines and 20 % for all other technologies. Boilers and CHP engines are subject to efficiency losses at part-load operation. In contrast, the COPs of turbo-driven compression and absorption chillers gain maximum values at 70 % and 55 % part-load, respectively.

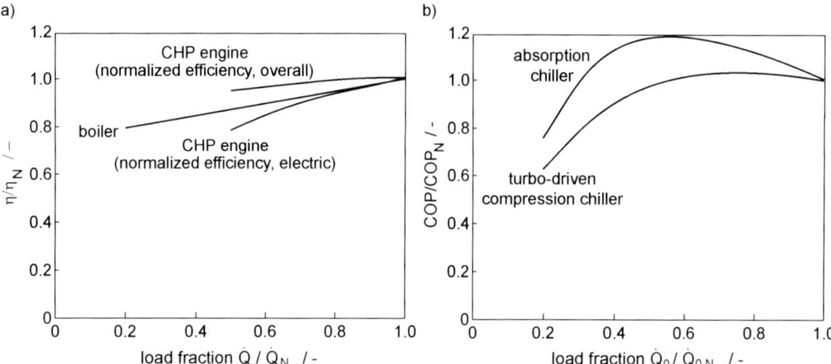

Figure 3.1: Characteristic part-load performance curves of heat generators (a) and chillers (b).

The model equations are listed in appendix A.1 representing the investment cost curves, the CHP engine's nominal electric and thermal efficiency curves, and the part-load performance curves.

3.2 An MILP formulation for DESS synthesis problems

For optimization-based synthesis of distributed energy supply systems, the *synthesis*, *design*, and *operation* level must be considered simultaneously (cf. section 2.1.2).

Hence, for single-objective optimization, the general mathematical programming problem is given by

$$\min_{s,d,o} f(s,d,o), \tag{3.2}$$
$$s \in S,\ d \in D,\ o \in O,$$

where the values of the decision variable vectors s, d, and o have to be determined to minimize some objective function f. The decision variables belong to the continuous and/or integer variable spaces S, D, and O, which represent the synthesis, design, and operation decision variable spaces, respectively.

In this section, an MILP formulation is presented that is based upon the work of Yokoyama et al. (2002). Quasi-stationary energy balances are assumed based on a discrete-time representation (Floudas and Lin, 2004; Stefansson et al., 2011) to account for multi-period demand profiles. Furthermore, constant quality levels are assumed, i.e., temperatures and pressures, at which the different energy forms are provided.

Objective function and balance equations

Kasas et al. (2011) identified the net present value (NPV) as best suited economic criterion for single-objective flowsheet optimization. Accordingly, the single objective function to be maximized is the net present value,

$$C_{t_{\mathrm{CF}}} = \frac{(i+1)^{t_{\mathrm{CF}}} - 1}{i \cdot (i+1)^{t_{\mathrm{CF}}}} \cdot R_{t_{\mathrm{CF}}} - I, \tag{3.3}$$

calculated from the net cash flow $R_{t_{\mathrm{CF}}}$ (annual revenues from feed-in electricity minus annual energy delivery and maintenance costs), total investments I, the cash flow time t_{CF}, and the discount rate i. For DESS, energy delivery costs are usually larger than feed-in revenues, and thus the net present value typically takes negative values.

In the following, the mathematical programming formulation is presented for a simple generic energy system (Fig. 3.2). The total system consists of up to n_{max} energy conversion units of which each unit n can either be existing ($y_n = 1$) or not ($y_n = 0$). For each instance of time t, a n with capacity $\dot{V}_{\mathrm{N,n}}$ converts input power (secondary energy, e.g., natural gas) $\dot{U}_{nt}$, purchased at price p_U, into output power (final energy, e.g., heat) $\dot{V}_{nt}$ to meet the time-varying energy demand $\dot{E}_t$ (e.g., heating, cooling or electricity). t_{max} discrete time steps with time step length Δt_t are assumed. The mathematical formulation of this optimization problem is as follows:

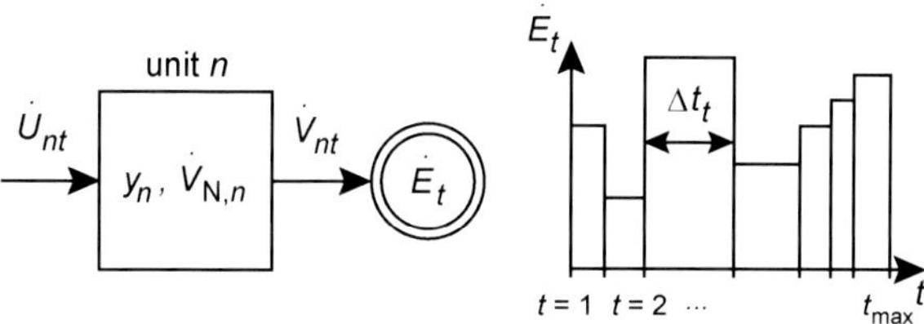

Figure 3.2: Flowsheet of a simple generic energy system. An energy conversion unit n with capacity $\dot{V}_{N,n}$ converts input power $\dot{U}_{nt}$ into output power $\dot{V}_{nt}$ to meet the time-varying energy demand $\dot{E}_t$ ($t_{\max}$ discrete time steps with time step length Δt_t). The (non-)existence of the conversion unit is represented by the binary decision variable y_n.

$$\max_{y_n, \dot{V}_{N,n}, \dot{V}_{nt}} \quad C_{t_{CF}} = -p_U \cdot \frac{(i+1)^{t_{CF}} - 1}{i \cdot (i+1)^{t_{CF}}} \cdot \sum_{n=1}^{n_{\max}} \sum_{t=1}^{t_{\max}} \Delta t_t \cdot \dot{U}_{nt}(\dot{V}_{nt}, \dot{V}_{N,n}) - \sum_{n=1}^{n_{\max}} I_n(y_n, \dot{V}_{N,n}), \tag{3.4}$$

$$\text{s.t.} \quad \sum_{n=1}^{n_{\max}} \dot{V}_{nt} = \dot{E}_t. \tag{3.5}$$

In this formulation, $\dot{V}_{nt}$ and $\dot{V}_{N,n}$ are continuous decision variables representing the operation (at time step t) and the continuous sizing of unit n, respectively. y_n is a binary decision variable representing the existence of unit n. Each of the technologies discussed in section 3.1 are represented by a component model composed of a cost function $I_n(y_n, \dot{V}_{N,n})$ and a performance function $\dot{U}_{nt}(\dot{V}_{nt}, \dot{V}_{N,n})$. This generic component-based modeling enables an easy model specification by establishing the energy balances for each energy demand (Eq. (3.5)).

For approximation of the nonlinear cost and performance functions, piecewise linear formulations are proposed. Misener et al. (2009) performed a comparative study of four different piecewise linear formulations. It is shown that the *linear segmentation method* as presented by Floudas (1995) is a computationally efficient linearization method. Therefore, in the following, the linear segmentation method is used for linearization of the investment cost curves and performance functions. For clarity, in the following discussion, the subscript n is omitted.

Investment cost

The investment cost functions presented in section 3.1 are approximated by piecewise linear functions according to the linear segmentation method. For the following discussion, two linear functions are assumed (Fig. 3.3).

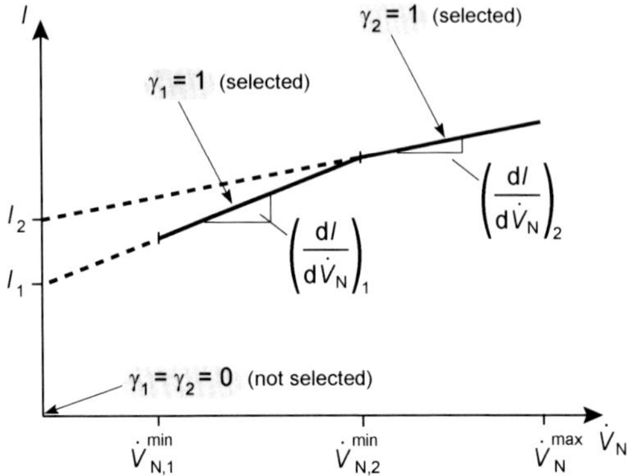

Figure 3.3: Exemplary piecewise linearized investment cost function. The binary decision variables γ_1 and γ_2 are highlighted in gray .

The following constraints model the investment cost for installing a piece of equipment

$$I(y, \gamma_1, \gamma_2, \dot{V}_{\mathrm{N}}) = \left(\gamma_1 \cdot I_1 + \dot{V}_{\mathrm{N}} \cdot \left(\frac{\mathrm{d}I}{\mathrm{d}\dot{V}_{\mathrm{N}}}\right)_1\right) + \left(\gamma_2 \cdot I_2 + \dot{V}_{\mathrm{N}} \cdot \left(\frac{\mathrm{d}I}{\mathrm{d}\dot{V}_{\mathrm{N}}}\right)_2\right), \tag{3.6}$$

$$y = \gamma_1 + \gamma_2 \leq 1, \tag{3.7}$$

$$\gamma_1 \cdot \dot{V}_{\mathrm{N},1}^{\min} \leq \dot{V}_{\mathrm{N}} \leq \gamma_1 \cdot \dot{V}_{\mathrm{N},2}^{\min}, \tag{3.8}$$

$$\gamma_2 \cdot \dot{V}_{\mathrm{N},2}^{\min} \leq \dot{V}_{\mathrm{N}} \leq \gamma_2 \cdot \dot{V}_{\mathrm{N}}^{\max}. \tag{3.9}$$

In Eqs. (3.6) to (3.9), γ_1 and γ_2 are binary decision variables representing the decision which piece of the cost curve is active. Eq. (3.7) makes sure that at most one piece of the linearized cost function is active at a time; moreover, it aggregates the binary decisions to the binary decision variable y representing the (non-)existence of the piece of equipment. Eqs. (3.8) and (3.9) guarantee that, if the unit is installed ($y = 1$), the nominal power $\dot{V}_{\mathrm{N}}$ lies either within the range between $\dot{V}_{\mathrm{N},1}^{\min}$ and $\dot{V}_{\mathrm{N},2}^{\min}$ ($\gamma_1 = 1$) or between $\dot{V}_{\mathrm{N},2}^{\min}$ and $\dot{V}_{\mathrm{N}}^{\max}$ ($\gamma_2 = 1$); or else, if the conversion unit is not installed ($y = 0$), the nominal power $\dot{V}_\mathrm{N}$ becomes zero. I_1 and I_2, and $\left(\frac{\mathrm{d}I}{\mathrm{d}\dot{V}_\mathrm{N}}\right)_1$ and $\left(\frac{\mathrm{d}I}{\mathrm{d}\dot{V}_\mathrm{N}}\right)_2$ are constant coefficients representing the axis intercepts and the slopes of the investment cost curves' piecewise linear approximation.

Operating performance

The operating efficiency $\eta_t(\dot{V}_t, \dot{V}_\mathrm{N})$, or for chillers the $\mathrm{COP}_t(\dot{V}_t, \dot{V}_\mathrm{N})$, relates a unit's input power $\dot{U}_t$ to its output power $\dot{V}_t$:

$$\dot{U}_t(\dot{V}_t, \dot{V}_\mathrm{N}) = \dot{V}_t / \eta_t(\dot{V}_t, \dot{V}_\mathrm{N}). \tag{3.10}$$

Usually, the efficiency curves $\eta_t(\dot{V}_t, \dot{V}_\mathrm{N})$ are nonlinear (cf. Fig. 3.1, section 3.1), and thus result in a nonlinear mathematical programming problem. However, these curves can be converted to $\dot{U}_t(\dot{V}_t)$-curves, which follow a much more linear trend. This way, even the part-load performance curve of an absorption chiller, which takes a maximum value at medium part-load, can be modeled adequately by only two piecewise linear functions (Fig. 3.4).

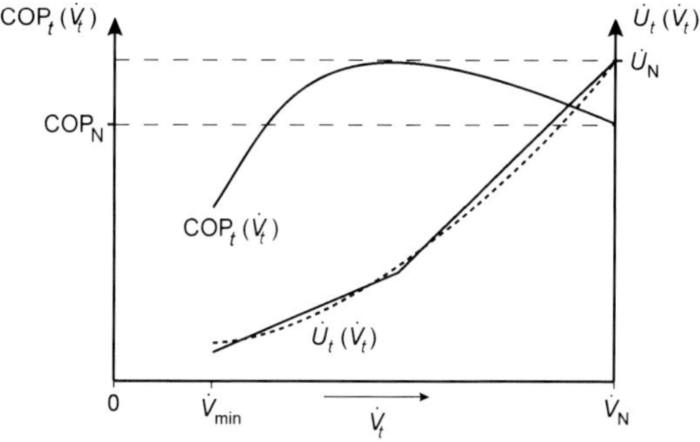

Figure 3.4: Part-load performance curves of an absorption chiller with capacity $\dot{V}_\mathrm{N}$ and minimum load $\dot{V}_\mathrm{min}$: Nonlinear $\mathrm{COP}_t(\dot{V}_t)$-curve, nonlinear $\dot{U}_t(\dot{V}_t)$-curve, and piecewise linearized $\dot{U}_t(\dot{V}_t)$-curve.

For simplicity of the presentation, in the following, the performance models are represented by one linear function only (Fig. 3.5):

$$\dot{U}_t(\delta_t, \dot{V}_t, \dot{V}_\mathrm{N}) = \delta_t \cdot \dot{U}_0(\dot{V}_\mathrm{N}) + \dot{V}_t \cdot \frac{\mathrm{d}\dot{U}_t}{\mathrm{d}\dot{V}_t}(\dot{V}_\mathrm{N}), \tag{3.11}$$

$$\delta_t \cdot \dot{V}_\mathrm{min}(\dot{V}_\mathrm{N}) \leq \dot{V}_t \leq \delta_t \cdot \dot{V}_\mathrm{N}. \tag{3.12}$$

In Eq. (3.11), δ_t is the binary decision variable representing the discrete on/off-status of the conversion unit at time step t. $\dot{U}_0(\dot{V}_\mathrm{N})$ and $\frac{\mathrm{d}\dot{U}_t}{\mathrm{d}\dot{V}_t}(\dot{V}_\mathrm{N})$ are coefficients representing the axis intercept and the slope of the performance model's linear approximation. Eq. (3.12) guarantees that, if the unit is in operation ($\delta_t = 1$), the supplied power $\dot{V}_t$ lies within the range of its minimum load $\dot{V}_\mathrm{min}(\dot{V}_\mathrm{N})$ and its capacity $\dot{V}_\mathrm{N}$; or, if the unit is not in operation ($\delta_t = 0$), the supplied power $\dot{V}_t$ becomes zero.

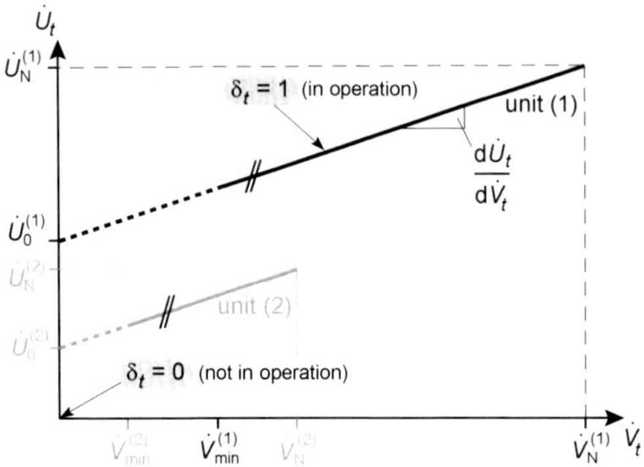

Figure 3.5: Exemplary linear performance functions of two units (1) and (2) with capacities $\dot{V}_\mathrm{N}^{(1)}$ and $\dot{V}_\mathrm{N}^{(2)}$, respectively. The binary decision variable δ_t represents the equipment's operation status, it is highlighted in gray.

In this performance model, the coefficients $\dot{U}_0(\dot{V}_\mathrm{N})$ and $\frac{\mathrm{d}\dot{U}_t}{\mathrm{d}\dot{V}_t}(\dot{V}_\mathrm{N})$ are still functions of the unit's capacity $\dot{V}_\mathrm{N}$. For continuous sizing, the characteristic performance models (cf. section 3.1) are used assuming that the part-load performance of a certain type of technology is valid for all units regardless of their capacities $\dot{V}_\mathrm{N}$. With this assumption, the coefficients in Eqs. (3.11) and (3.12) can be expressed as functions of the nominal capacities $\dot{V}_\mathrm{N}$ as follows (for illustration, compare Fig. 3.5):

$$\dot{U}_0(\dot{V}_\mathrm{N}) = u_0 \cdot \dot{U}_\mathrm{N} = u_0 \cdot \frac{\dot{V}_\mathrm{N}}{\eta_\mathrm{N}}, \tag{3.13}$$

$$\frac{\mathrm{d}\dot{U}_t}{\mathrm{d}\dot{V}_t}(\dot{V}_\mathrm{N}) \overset{\text{linear}}{=} \frac{\mathrm{d}\frac{\dot{U}_t}{\dot{U}_\mathrm{N}}}{\mathrm{d}\frac{\dot{V}_t}{\dot{V}_\mathrm{N}}} \cdot \frac{\dot{U}_\mathrm{N}}{\dot{V}_\mathrm{N}} = \frac{\mathrm{d}u_t}{\mathrm{d}v_t} \cdot \frac{1}{\eta_\mathrm{N}}, \tag{3.14}$$

$$\dot{V}_\mathrm{min}(\dot{V}_\mathrm{N}) = \dot{V}_\mathrm{N} \cdot v_\mathrm{min}. \tag{3.15}$$

In Eqs. (3.13) and (3.14), u_0 and $\frac{\mathrm{d}u_t}{\mathrm{d}v_t}$ are constant coefficients representing the intercept and the slope of the linear approximation of the characteristic performance model (cf. section 3.1). For convenience, the unit's input power $\dot{U}_\mathrm{N}$ is expressed in terms of its output power $\dot{V}_\mathrm{N}$ and its nominal efficiency η_N, which is assumed to be constant over the range of equipment sizes. In Eq. (3.15), v_min is a constant coefficient representing the minimum relative input power, i.e.,the minimum load fraction, of the unit. In summary, the characteristic performance model is given by:

$$\dot{U}_t(\delta_t, \dot{V}_t, \dot{V}_\mathrm{N}) = \delta_t \cdot u_0 \cdot \frac{\dot{V}_\mathrm{N}}{\eta_\mathrm{N}} + \dot{V}_t \cdot \frac{\mathrm{d}u_t}{\mathrm{d}v_t} \cdot \frac{1}{\eta_\mathrm{N}}, \tag{3.16}$$

$$\delta_t \cdot \dot{V}_\mathrm{N} \cdot v_\mathsf{min} \leq \dot{V}_t \leq \delta_t \cdot \dot{V}_\mathrm{N}. \tag{3.17}$$

Linearization of the performance model

In Eqs. (3.16) and (3.17), the term $\delta_t \cdot \dot{V}_\mathrm{N}$ is a bilinear product of a binary and a continuous decision variable as analyzed by Rodriguez and Vecchietti (2013). The bilinear term represents a nonlinearity for the optimization problem that is circumvented by applying the reformulation strategy developed by Petersen (1971) and Glover (1975). This reformulation strategy exploits the linear equivalent of a bilinear term and can be applied to bilinear products of two binary, or one binary and one continuous decision variable: The bilinear product is substituted by a single continuous decision variable whose value is determined by linear constraints guaranteeing that the behavior of the bilinear product is reproduced correctly. In the context of process systems engineering, this approach was first employed by Psarris and Floudas (1990). Yokoyama et al. (2002)[1] used this reformulation strategy for the optimization-based synthesis of DESS.

For the presented formulation, the bilinear product $\delta_t \cdot \dot{V}_\mathrm{N}$ is substituted by a continuous auxiliary variable ξ_t. However, other than δ_t, the capacity $\dot{V}_\mathrm{N}$ is not time-dependent, and thus a detour becomes necessary: The continuous variable Ψ_t is introduced that takes the value of $\dot{V}_\mathrm{N}$ for all time steps t,

$$\Psi_t = \dot{V}_\mathrm{N} \quad \forall t. \tag{3.18}$$

Next, in Eqs. (3.4)-(3.17), all occurrences of $\dot{V}_\mathrm{N}$ are replaced by Ψ_t, and thus the bilinear product $\delta_t \cdot \dot{V}_\mathrm{N}$ becomes $\delta_t \cdot \Psi_t$, and can eventually be substituted by ξ_t. The following linear constraints are added,

[1]Please note that there has been a typo in (Yokoyama et al., 2002): On page 777 in the explanatory text following Eq. (9) it should be $\xi_{ij} = 0$ instead of $\zeta_{ij} = 0$.

$$\delta_t \cdot V_{\mathrm{N}}^{\min} \leq \xi_t \leq \delta_t \cdot V_{\mathrm{N}}^{\max}, \tag{3.19}$$

$$(1-\delta_t) \cdot \dot{V}_{\mathrm{N}}^{\min} \leq \Psi_t - \xi_t \leq (1-\delta_t) \cdot \dot{V}_{\mathrm{N}}^{\max}, \tag{3.20}$$

guaranteeing that ξ_t correctly reproduces the behavior of the substituted bilinear product $\delta_t \cdot \dot{V}_{\mathrm{N}}$, that is:

- If $\delta_t = 0$, then $\xi_t = \delta_t \cdot \dot{V}_{\mathrm{N}} = 0$ with $\dot{V}_{\mathrm{N}}^{\min} \leq \dot{V}_{\mathrm{N}} \leq \dot{V}_{\mathrm{N}}^{\max}$, and
- if $\delta_t = 1$, then $\Psi_t = \xi_t = \dot{V}_{\mathrm{N}}$ with $\dot{V}_{\mathrm{N}}^{\min} \leq \dot{V}_{\mathrm{N}} \leq \dot{V}_{\mathrm{N}}^{\max}$.

The derived MILP formulation enables to simultaneously optimize the configuration, sizing and operation of DESS. It accounts for time-varying load profiles, continuous equipment sizing, and part-load dependent operating efficiencies. Thus, the presented approach is capable of assessing the major trade-offs inherent to DESS synthesis problems using robust MILP optimization algorithms.

3.2.1 Generalization of the MILP formulation

To use the presented MILP formulation for practical DESS synthesis problems, some straightforward extensions are necessary. First of all, the generalized formulation takes into account more than one linear function to approximate the characteristic performance model. The model equations representing the piecewise linearized equipment models employed in this thesis are listed in appendix A.2. In addition, according to Table 3.1, maintenance cost are considered as constant factors in terms of equipment investment. Moreover, the formulation considers multiple energy demands, and multiple input and output powers as required to model polygeneration units like CHP engines. Furthermore, the modeling is extended to include units like absorption chillers that draw final energy supplied by some other conversion unit (heat) to produce a different final energy form (cooling).

Finally, to capture the dependency of the CHP engine's nominal electric and thermal efficiencies on the engine's size (cf. section 3.1), the CHP model is partitioned into three size classes with constant, but for each size class different, efficiencies (cf. appendix A.1):

- Small: 500 ... 1400 kW, $\eta_{\mathrm{N,th}} = 0.409$, $\eta_{\mathrm{N,el}} = 0.462$,
- Medium: 1400 ... 2300 kW, $\eta_{\mathrm{N,th}} = 0.434$, $\eta_{\mathrm{N,el}} = 0.435$,
- Large: 2300 ... 3200 kW, $\eta_{\mathrm{N,th}} = 0.463$, $\eta_{\mathrm{N,el}} = 0.435$.

Within the CHP model, each size class is represented by a component model of its own. To guarantee that, for each CHP unit in the superstructure, at most one of these size classes can be active at a time, the following constraint is added to the CHP engine model:

$$y_{\text{Small}} + y_{\text{Medium}} + y_{\text{Large}} \leq y, \tag{3.21}$$

where y_{Small}, y_{Medium}, and y_{Large} are binary decision variables representing the existence of each size class of a CHP unit, whose existence is represented by the binary decision variable y.

Simplifying assumptions

The presented MILP formulation neglects energy storages and the layout of the energy distribution network. Moreover, the component set is limited to boilers, CHP engines, compression chillers, and absorption chillers. The employed MILP formulation can easily be extended to account for energy storage devices, the network layout, and further equipment models (see section 7.1). However, these extensions will increase the complexity of the synthesis problems without contributing to the meaningfulness of the synthesis methods proposed in this thesis. Therefore, these extensions are avoided in this thesis.

3.2.2 Combinatorial complexity of DESS synthesis problems

Integer constraints to avoid combinatorial redundancy

The combinatorial complexity of a superstructure optimization problem increases exponentially with the number of units considered in the superstructure: For each unit, there is one discrete decision, i.e., two choices, for equipment selection (on/off). Thus, a superstructure embedding n units, incorporates 2^n different solutions structures. However, a significant number of these solutions represents combinatorially redundant structures, i.e., different integer solutions that represent identical structures. Consider, e.g., a superstructure incorporating three boilers, which can be either existent, "1", or non-existent, "0". The set of combinations is $\{\{0,1\},\{0,1\},\{0,1\}\}$. However, e.g., the structures $\{1,0,0\}$, $\{0,1,0\}$, and $\{0,0,1\}$ are combinatorially redundant.

To avoid combinatorial redundancy in the search space, logical constraints are introduced into the MILP formulation to limit the combinations for equipment selection to combinatorially non-redundant alternatives. Consider N units of the same type of technology embedded in the superstructure: The (non-)existence of each unit $n \in N$

is represented by the binary decision variable y_n. The following constraints,

$$y_n \leq y_{n-1}, \quad \forall\, i = 2, \ldots, N, \tag{3.22}$$

ensure that unit n can be selected only after unit $(n-1)$ has already been selected. If the synthesis problem incorporates more than one energy demand, these constraints are added for each unit of the same type of technology and for each energy demand.

Number of solution structures embedded in superstructure models

As a measure for the complexity of a particular superstructure-based synthesis problem, the number of solution structures embedded in the superstructure model is assumed. In general, a superstructure model incorporating n units embeds 2^n different solutions structures. However, this number does not yet reflect that combinatorially redundant solutions are avoided (i), and that the CHP engine model is partitioned into three size classes (ii).

i) Avoided combinatorial redundancy. The constraint (3.22) avoids combinatorial redundancy. In combinatorics, this is called a *k-combination with repetitions*, in which order is not taken into account (Grimaldi, 1998). The number of combinations to choose k elements (here, k represents units of the same type of technology connected to one and the same energy demand) from a set of n elements (number of choices per unit; here, two: "selected" or "not selected") allowing for duplicates (i.e., identical choices for different elements; e.g., two boilers can be both selected or not), but disregarding different orderings (to avoid combinatorial redundancy), is given by the binomial coefficient

$$\binom{n+k-1}{k} = \frac{(n+k-1)!}{(n-1)! \cdot k!} \tag{3.23}$$

ii) Partitioning of CHP engine model. The partitioning of the CHP engine model into three size-classes increases the combinatorial complexity, because, for each CHP engine, there are four choices: no CHP engine, one small-, one medium-, or one large-sized CHP engine. Thus, the number of combinations to arrange k CHP engines encoded in the superstructure model are calculated by the binomial coefficient

$$\binom{n+k-1}{k} \quad \text{with } n = 4.$$

In summary, the number of solution structures embedded in a superstructure model based upon the presented MILP formulation is given by

$$\prod_{\substack{\text{for each type of technology} \\ \text{connected to one and the} \\ \text{same energy demand}}} \binom{n+k-1}{k} \cdot \quad \dots \quad \cdot \prod_{\text{CHP engines}} \binom{4+k-1}{k}.$$

Keep in mind, that this formula takes into account only the discrete decisions due to equipment selection. The complete synthesis problem is substantially more complex due to the additional decisions for equipment sizing and operation.

3.3 Evaluation of the MILP formulation

Intrinsically, optimization-based DESS synthesis is an MINLP problem due to the existence of both continuous and discrete decision variables and their nonlinear relationships in the objective function and the constraints (cf. section 3.3.1). Hence, the proposed MILP formulation is only a linear approximation of the original MINLP problem. In this section, the MILP formulation is evaluated with regard to its approximation of the original MINLP formulation and computational effort to solve a synthesis test problem.

3.3.1 Original MINLP formulation

In analogy to the linearized MILP formulation (Eqs. (3.4)-(3.20)), the original MINLP formulation is given by

$$\begin{aligned}
\max_{y_n, \dot{V}_{\mathrm{N},n}, \delta_{nt}, \dot{V}_{nt}} \quad C_{t_{\mathrm{CF}}} = & - \; p_U \cdot \frac{(i+1)^{t_{\mathrm{CF}}} - 1}{i \cdot (i+1)^{t_{\mathrm{CF}}}} \cdot \sum_{n=1}^{n_{\max}} \sum_{t=1}^{t_{\max}} \Delta t_t \cdot \dot{U}_{nt}(\delta_{nt}, \dot{V}_{nt}, \dot{V}_{\mathrm{N},n}) \\
& - \sum_{n=1}^{n_{\max}} I_n(y_n, \dot{V}_{\mathrm{N},n}), \\
\text{s.t.} \quad & \sum_{n=1}^{n_{\max}} \dot{V}_{nt} = \dot{E}_t, \\
& y_n \cdot \dot{V}_{\mathrm{N},n}^{\min} \leq \dot{V}_{\mathrm{N},n} \leq y_n \cdot \dot{V}_{\mathrm{N},n}^{\max}, \\
& \delta_{nt} \cdot \dot{V}_{\min}(\dot{V}_{\mathrm{N},n}) \leq \dot{V}_{nt} \leq \delta_t \cdot \dot{V}_{\mathrm{N},n},
\end{aligned}$$

in which $I_n(y_n, \dot{V}_{\mathrm{N},n})$ and $\dot{U}_{nt}(\delta_{nt}, \dot{V}_{nt}, \dot{V}_{\mathrm{N},n})$ are the discrete-continuous nonlinear investment cost and performance functions, respectively. y_n and δ_{nt} are binary decision variables representing the (non-)existence and the discrete on/off-status of equipment n at time step t, respectively. $\dot{V}_{\mathrm{N},n}$ and $\dot{V}_{nt}$ are continuous decision variables representing the sizing and operation of unit n at time step t, respectively. The objective function models the net present value for cash flow time t_{CF} and discount rate i.

3.3.2 Synthesis test problem

A simple grassroots synthesis test problem is considered comprising time-varying heating and cooling demands (Table 3.2). The considered site is connected to the regional natural gas grid (gas tariff: 6 ct/kWh) and the regional electricity grid (electricity tariff: 16 ct/kWh; feed-in tariff: 10 ct/kWh). For net present value calculations, a cash flow time of 10 years and a discount rate of 8 % are assumed.

Table 3.2: Seasonal-averaged demands for heating and cooling including summer and winter peak-loads. The peak-loads occur only during few hours per year, and thus hardly contribute to the annual energy demands.

	Winter (Peak)	Spring	Summer (Peak)	Fall
Heating / MW	2.4 (4.3)	1.5	0.7	1.5
Cooling / MW	1.2	1.3	2.6 (3.1)	1.9

For the solution of the synthesis problem, the component set introduced in section 3.1 is used (i.e., boilers, CHP engines, compression chillers, and absorption chillers). A superstructure is assumed that incorporates two units of each technology (Fig. 3.6). Electricity produced by the CHP engines can be fed-in to the regional electricity grid, or it can be directly used on-site to power the turbo-chillers.

3.3.3 Combinatorial complexity of the test problem

The superstructure model encodes 270 possible solution structures (cf. section 3.2.2):

$$\underbrace{\binom{2+2-1}{2}}_{\text{boilers}} \cdot \underbrace{\binom{4+2-1}{2}}_{\text{CHP engines}} \cdot \underbrace{\binom{2+2-1}{2}}_{\text{turbo-chillers}} \cdot \underbrace{\binom{2+2-1}{2}}_{\text{absorption chillers}} = 3 \cdot 10 \cdot 3 \cdot 3 = 270.$$

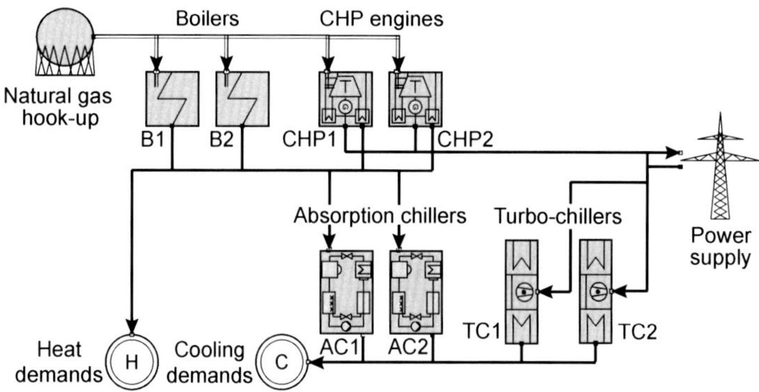

Figure 3.6: Test problem superstructure incorporating two boilers (B1 and B2), two CHP engines (CHP1 and CHP2), two turbo-driven compression chillers (TC1 and TC2), and two absorption chillers (AC1 and AC2).

A number of 270 possible solution structures seems to be fairly small. However, keep in mind that the complete synthesis problem is significantly more complex due to the additional decisions for equipment sizing and operation. In particular, the operation optimization considerably complicates the synthesis problem due to the additional decisions for the units' discrete on/off-status and the units' operating points. Note that the operational decisions appear in each time step. In fact, even for the simple test case, the resulting MILP and MINLP problems incorporate already quite a large number of variables and constraints (cf. Table 3.3). Compared to the MINLP problem, the problem size of the MILP problem is almost sixfold in terms of constraints and variables due to the introduction of binary decision variables for the linearization of the component models.

Table 3.3: Test problem sizes for MILP and MINLP formulations.

	MILP problem	MINLP problem
constraints	4116	712
continuous variables	3030	506
binary variables	328	56

3.3.4 Solution of MILP and MINLP test problems

The resulting MILP and MINLP problems are implemented in the modeling language GAMS version 23.7.3 (Brooke et al., 2010). For the solution of the MILP problem, IBM ILOG CPLEX version 12.2 (IBM ILOG, 2011) is used. For the solution of the MINLP problem, the MINLP solver COIN-OR Bonmin version 1.4 (Bonami et al., 2007) and the global optimizer BARON version 9.3.1 (Tawarmalani and Sahinidis, 2004, 2005) are used. For the solution of the NLP subproblems, Ipopt version 3.8 (Wächter and Biegler, 2006) is used. The computational tests are performed on an Intel Xeon X5650 2.67 GHz with 2 GB RAM and four kernels. For better comparison, only one kernel is used. The operation system is Windows XP (32-bit). For all solvers, the default settings are used. The problems are solved to global optimality (maximal relative optimality gap: 0 %). A time limit of 48 hours is specified, after which the computation is aborted.

The optimization results are listed in Table 3.4. For the MILP problem, no initial solution is provided. For the MINLP problem, a *trivial* initial solution is provided. In the trivial initial solution, all units are initialized with maximum capacities and maximum power flows for all time steps. The MILP problem is solved within 88 seconds employing three CHP partitions and at maximum three nodes for piecewise linearization of the cost and performance curves (cf. appendix A.2). The MINLP-solvers are not able to solve the problem to optimality before the time limit of 48 hours is reached. If the MILP solution is provided as an initial solution for the MINLP problem, Bonmin still only finds a suboptimal solution. On the other hand, initialized with the MILP solution, BARON solves the MINLP problem to global optimality. However, the computation time of the global solver BARON amounts to over 23 hours. A detailed list of the equipment installed in the optimal MILP and MINLP solutions including nominal thermal powers, investment costs, operating times, and annual average part-loads is given in appendix C.1.

The solution obtained from the linearized MILP problem is structurally identical with the global optimal MINLP solution, but slightly differs with respect to equipment sizing; then again, the operation strategies of both solutions resemble each other very much (cf. appendix C.1). Furthermore, if the MILP solution is fixed in the MINLP model (structure, sizing, and operation), evaluation of the MINLP model yields an NPV of −7.14 M€, which equals a relative difference from the global optimal MINLP objective function value of only 0.45 %. This difference can be mainly attributed to deviations between the original nonlinear and the linearized performance models.

Table 3.4: Comparison of computations using the MILP and MINLP formulations for the synthesis test problem. Objective function: net present value (NPV). CPU times given in h:mm:ss (t_∞: time limit of 48 hours reached, optimization aborted). Maximal relative optimality gap: 0 %.

	NPV / M€	CPU time	Solution structure (equipment sizing in kW) B1	B2	CHP1	CHP2	TC1	TC2	AC1	AC2
MILP										
CPLEX	−7.24	0:01:28	1900	100	2300	0	1900	840	370	0
MINLP (trivial initial solution)										
Bonmin	−11.76	1:41:15	2160	100	3200	800	1170	0	3380	0
BARON		t_∞								
MINLP (initialized with MILP solution)										
Bonmin	−10.96	0:16:49	650	0	3200	1110	2820	0	2660	0
BARON	−7.11	23:12:11	1900	280	2120	0	1900	800	410	0

If both a finer partitioning of the CHP model and a finer discretization of the piecewise linearized cost and performance curves are used, the solution structures of the corresponding MILP problems remain unchanged (Table 3.5). Moreover, the MILP sizing converges to the global optimal MINLP solution, however, at the expense of drastically increasing solution times. In fact, with six CHP partitions and six nodes for the piecewise linearization, the MILP solution is identical[2] with the global optimal MINLP solution, yet the computation takes 55:57 minutes (compared to 88 seconds (2.6 %), if only three CHP partitions and three nodes for the piecewise linearization are employed).

While a formal validation of the presented modeling framework is beyond the scope of the present work, it should be noted that the discussed test problem is representative for general DESS synthesis problems: The special characteristics inherent to DESS synthesis problems are reflected within the test problem (economy of scale for equipment investments, limited equipment capacities, part-load performance, minimum operating loads, and multiple redundant units). Moreover, the component set encompasses simple heating and cooling equipment (boilers and compression chillers) as well as thermally-driven chillers (absorption chillers) and polygeneration units (CHP engines).

[2]Note that the equipment sizing is rounded to the tens.

Table 3.5: Influence of partitioning and discretization of the linearized model on solution quality and solution time: Number of nodes for piecewise linearization n_n, number of CHP model partition n_p. Equipment sizing rounded to the tens.

$n_n = n_p$	NPV / M€	CPU time	Solution structure (equipment sizing in kW) B1	B2	CHP1	CHP2	TC1	TC2	AC1
MINLP	−7.11	23:12:11	1900	280	2120	-	1900	800	410
MILP									
2	−7.53	0:00:27	1900	0	1500	900	1900	1210	0
3	−7.24	0:01:28	1900	100	2300	0	1900	840	370
4	−7.18	0:03:42	1860	100	2340	0	1900	860	350
5	−7.11	0:13:36	1900	250	2150	0	1900	820	390
6	−7.11	0:55:57	1900	280	2120	0	1900	800	410

In summary, even for this simple example and even when a good initial solution is provided, a robust solution of the full MINLP problem cannot be guaranteed. It should be noted again, however, that, for all solvers, only the default settings were used. In contrast, the solution of the MILP problem takes only a few seconds if at most two curves are assumed for the piecewise linearization of the cost and performance curves, and if three partitions are assumed for the CHP engine model ($n_p = n_n = 3$). Moreover, with these equipment models, the MILP solution is structurally identical with the global optimal MINLP solution. Since the focus of this thesis is on algorithm development rather than on modeling, in the remainder of this thesis, the MILP formulation is employed with $n_p = n_n = 3$.

3.4 Summary and conclusions

In this chapter, an MILP framework is proposed for the synthesis of distributed energy supply systems (DESS). First, a set of component models is introduced encompassing boilers, CHP engines, turbo-driven compression chillers, and absorption chillers. Each component model is composed of an investment cost function and a performance curve. Based on the component-based modeling framework, a mathematical programming formulation is presented. Intrinsically, optimal DESS synthesis is an MINLP problem due to the existence of both continuous and discrete decision variables and their nonlinear relationships in the objective function and the constraints. However, for robust optimization, an MILP formulation is derived that enables to rigorously optimize the structure, sizing, and operation of distributed energy supply systems represented by superstructure models that account for time-varying load profiles, continuous equipment sizing, and part-load dependent operating efficiencies. Furthermore, to limit the combinatorial complexity of the optimal synthesis problem, the proposed MILP formulation avoids combinatorial redundancy, i.e., different integer solutions that represent identical structures. Finally, it is shown that the MILP formulation adequately approximates the original MINLP formulation while enabling fast and robust optimization. In the remainder of this thesis, the presented MILP formulation is employed for the optimal synthesis of distributed energy supply systems.

Chapter 4

Automated superstructure-based synthesis of distributed energy supply systems

In this chapter, an automated superstructure-based synthesis framework is proposed for the optimal synthesis of distributed energy supply systems (DESS). The presented approach fulfills the requirements for optimal DESS synthesis as discussed in section 2.5.

Major parts of this chapter are based on earlier publication by Voll et al. (2013). The proposed methodology consists of two parts: an algorithm for automated generation of DESS superstructures (section 4.1), and a successive optimization algorithm that continuously increases the number of units included in the superstructure (section 4.2). In section 4.3, the proposed methodology is applied to an illustrative synthesis problem identifying the optimal solution and further near-optimal solution alternatives. A short summary concludes this chapter (section 4.4).

4.1 Automated superstructure and model generation

The proposed algorithm for automated superstructure and model generation (Fig. 4.1) relies on a basic problem definition that includes the following information: demand time series, specification of existing equipment as well as available new equipment, and, optionally, topographic constraints. In addition, the user can specify the number of redundant units to be considered in the superstructure. If no information is available on the number of redundant units, the algorithm automatically assumes one unit of each technology.

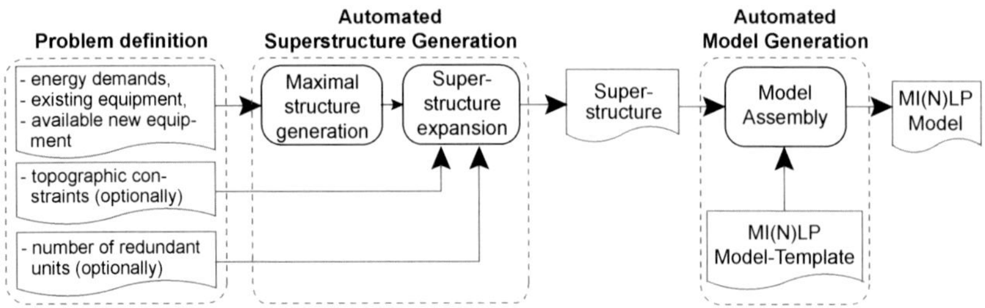

Figure 4.1: Flow diagram of the algorithm for automated superstructure and model generation.

As a first step, the algorithm employs the P-graph based *maximal structure generation* (MSG) algorithm to generate a superstructure including exactly one unit of each *plausible* technology, i.e., all technologies that can be reasonably connected to supply the required energy forms (for more details, see section 2.3.2). Next, a matrix representation is constructed to represent the superstructure as a *connectivity matrix*. This matrix is used to expand the superstructure to incorporate multiple redundant units and user-defined topographic constraints. The expanded connectivity matrix can directly be used to assemble the final optimization model employing user-defined MI(N)LP model-templates. This generic modeling framework allows to use arbitrary mathematical programming formulations. The presented framework is implemented in Java (Bloch, 2008).

4.1.1 Maximal structure generation

The application of the maximal structure generation (MSG) algorithm is illustrated for the synthesis of a simple heating and cooling system (Fig. 4.2). For each conversion technology, the input and output connections are defined by the energy carriers that are transferred along these connections, i.e., hot and cold water. For simplicity, in the following discussion, electricity and natural gas supply are neglected. The problem description is shown in Fig. 4.2 a): Boilers, absorption and turbo-driven compression chillers are available to meet the requirements of one heating and two cooling demands. The MSG algorithm employs the definition of the input and output energy carriers to connect compatible conversion units with each other, i.e., output connections are connected with input connections if they transfer the same type of energy. Thereby, the algorithm automatically generates the maximal structure (Fig. 4.2 b) incorporating one boiler, one absorption chiller, and one turbo-chiller. The boiler supplies hot water to the heat demand and to the absorption chiller. Both chillers supply cold water to the two cooling demands.

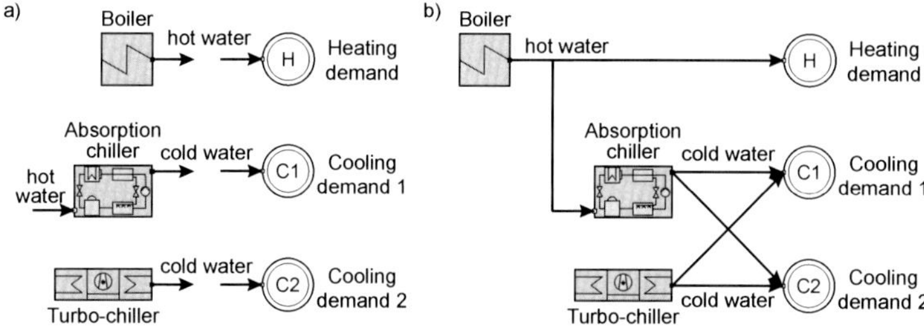

Figure 4.2: Application of the MSG algorithm for the synthesis of a simple heating and cooling system. For simplicity, electricity and natural gas supply are not shown. a) Problem definition with energy demands, energy carriers, and available energy conversion units. b) Maximal structure.

4.1.2 Connectivity matrix representation

To incorporate redundant conversion units into the generated superstructure, a connectivity matrix representation is used: Rows and columns of connectivity matrix **C** represent final energy users (e.g., demands for heating and cooling) and generators, such as heating generators (e.g., boilers and CHP engines) and cooling generators (e.g., absorption and compression chillers), respectively. The entries c_{lk} of the matrix **C** represent the connectivity between generators and users: If a generator l is connected to a user k, the corresponding entry is $c_{lk} = 1$; otherwise, it is 0. In Fig. 4.3, the connectivity matrix is shown that corresponds to the maximal structure from Fig. 4.2 b). The absorption chiller takes a special role since it represents both a generator (cooling) and a final energy user (driving heat).

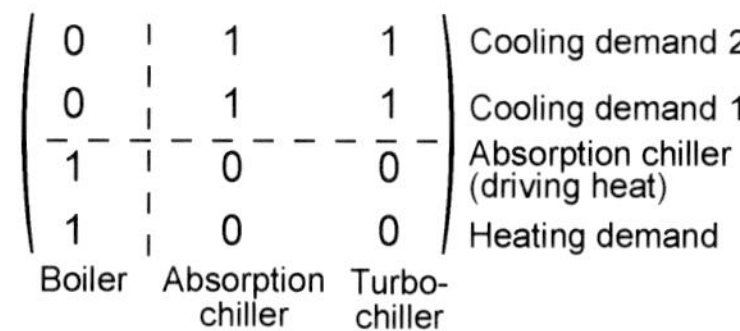

Figure 4.3: Connectivity matrix of the maximal structure incorporating one heating and two cooling demands supplied by one boiler, one absorption chiller, and one turbo-chiller (cf. Fig. 4.2 b). For simplicity, electricity and natural gas are not shown.

4.1.3 Superstructure expansion

Superstructure expansion, i.e., incorporation of multiple redundant units and topographic constraints, is performed by manipulation of the connectivity matrix. The incorporation of topographic constraints is the more complex of the two expansion tasks, and thus incorporation of multiple redundant units is presented first. When adding a redundant unit to the superstructure, the connectivity of the redundant unit is identical to the already existing unit of the same technology. Or more technically, redundant units are generated by adding copies of already existing rows and columns to the connectivity matrix: To incorporate n redundant units of a particular technology in the superstructure, all columns and rows representing this technology are copied n-times.

In Fig. 4.4, the example from Fig. 4.2 is expanded by a second redundant absorption chiller. For this purpose, the column representing the cooling supply of the already existing absorption chiller AC1 is copied. Next, the row representing the driving heat demand of the absorption chiller AC1 is copied. The generated connectivity matrix (Fig. 4.4 b) now represents the extended superstructure (Fig. 4.4 a). Finally, the mathematical programming model is specified by establishing the energy balances for each energy demand according to the generated connectivity matrix. In this thesis, the MILP framework presented in chapter 3 is used.

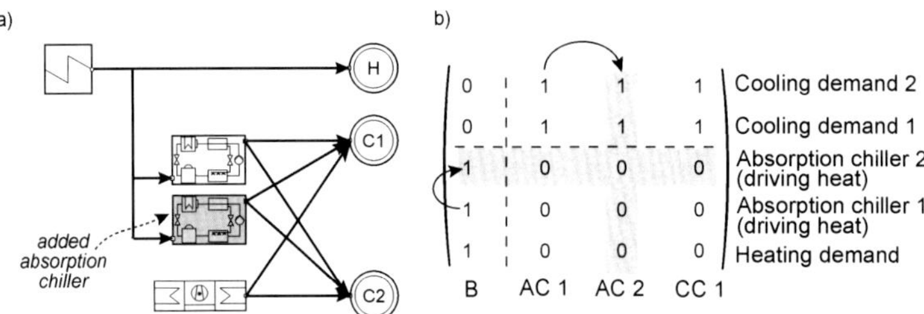

Figure 4.4: a) Expanded superstructure incorporating two absorption chillers. b) Expanded connectivity matrix. Copied columns and rows are highlighted in gray. B: boiler, AC: absorption chiller, CC: compression chiller.

Consider the example from Fig. 4.2 once more. This time, it is expanded by adding a topographic constraint that enforces decentralized cooling supplies. Thus, this constraint prohibits that one and the same chiller unit simultaneously satisfies all cooling demands. Accordingly, the corresponding connections in the superstructure cannot be used at the same time, and thus have to be cut (Fig. 4.5 a). Still, for the optimal synthesis, the installation of both chiller types for both cooling demands has to be evaluated. For this purpose, the superstructure is expanded by adding the missing units for each cooling demand (Fig. 4.5 b). In the corresponding connectivity matrix, the full expansion is realized by assigning zeros to the entries representing the deleted connections, and by adding one column for each added generator (here, one column for the added compression chiller and one column for the added absorption chiller) and one row for each added energy user (here, one row for the driving heat demand of the added absorption chiller) to the connectivity matrix (Fig. 4.5 c).

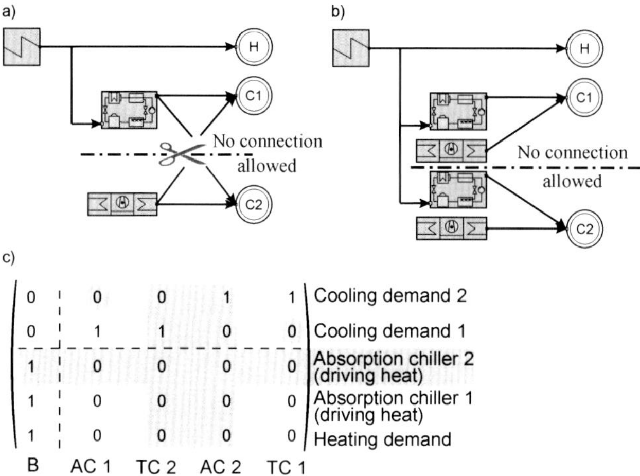

Figure 4.5: a) Flowsheet representation of a superstructure accounting for the topographic constraint enforcing decentralized cooling supplies of two cooling demands. b) Expanded superstructure incorporating one turbo-compression and one absorption chiller for each cooling demand. c) Connectivity matrix of the expanded superstructure. Changes are highlighted in gray. For simplicity, electricity and natural gas are not shown. B: boiler, AC: absorption chiller, CC: compression chiller.

4.1.4 Generalization to integrate further energy forms

The preceding examples are limited to superstructure expansions for the inclusion of multiple heating and cooling generators. It should be noted though that the proposed algorithm is not restricted to these units. In fact, the algorithm can directly be applied to include energy conversion units providing other energy forms.

In this work, the following energy carriers are considered: natural gas, electricity, hot water, and cold water. Accordingly, energy demands for electricity, heating, and cooling are included. The extension by further energy carriers (and discrete quality levels) is straightforward. In Fig. 4.6, the maximal structure and the corresponding connectivity matrix are provided for an example, in which all conversion technologies and energy carriers considered within this work are explicitly shown. The maximal structure incorporates one boiler, one CHP engine, one absorption chiller, and one turbo-driven compression chiller supplying hot water, cold water, and electricity to one heating, one cooling, and one electricity demand. The boiler and CHP engine use natural gas delivered by the natural gas hook-up. The turbo-chiller employs electricity, which is partly supplied by the power supply, and partly by the CHP engine. The absorption chiller is driven with heat that is provided by the boiler and the CHP engine. All CHP electricity that is not used on-site is fed-in to the public energy market represented by the power supply. Apparently, by explicit consideration of electricity and gas, all energy conversion units generally are both generators and users due the need for driving power; e.g., a boiler requires natural gas to generate heating; and the power supply generates electricity, but also absorbs feed-in electricity. Thus, energy conversion units appear in both columns and rows of the connectivity matrix (Fig. 4.6 b). For the sake of simplicity, in the remainder of this thesis, electricity and natural gas are not shown in the connectivity matrices.

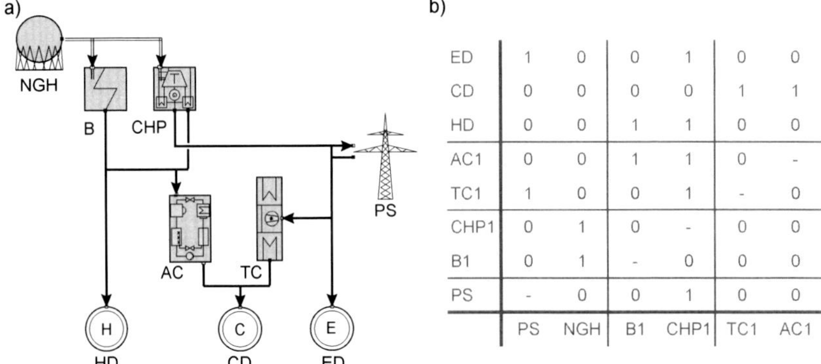

ED	1	0	0	1	0	0
CD	0	0	0	0	1	1
HD	0	0	1	1	0	0
AC1	0	0	1	1	0	-
TC1	1	0	0	1	-	0
CHP1	0	1	0	-	0	0
B1	0	1	-	0	0	0
PS	-	0	0	1	0	0
	PS	NGH	B1	CHP1	TC1	AC1

Figure 4.6: Maximal structure considering hot and cold water as well as electricity and natural gas. a) Flowsheet representation. b) Connectivity matrix representation. B: boiler, CHP: CHP engine, AC: absorption chiller, TC: turbo-driven compression chiller, HD/CD/ED: heating/cooling/electricity demand, NGH: natural gas hook-up, PS: power supply.

4.2 Successive superstructure generation and optimization

The algorithm for superstructure and model generation automatically generates a mathematical programming model for a given synthesis problem. Furthermore, it automatically incorporates topographic constraints and a given number of multiple redundant units into the problem formulation. However, when solving DESS synthesis problems, a major difficulty is that the designer does not know *a priori* how many redundant units have to be included in the superstructure to guarantee the inclusion of the optimal solution. Then again, if too many units are included in the model, the computational effort to solve this problem becomes prohibitively large. To address this problem, a successive optimization approach is proposed that continuously increases the number of units included in the superstructure until no further improvement is observed (Fig. 4.7) .

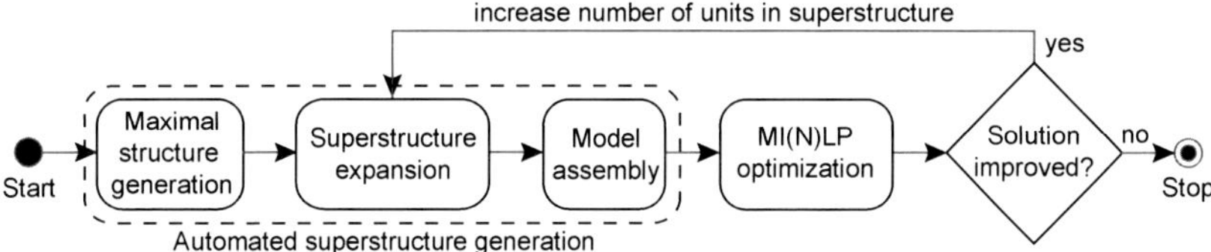

Figure 4.7: Flow diagram representing the successive algorithm for automated superstructure generation and optimization of DESS synthesis problems.

First, the successive algorithm employs the automated superstructure and model generation algorithm to generate and expand the maximal structure, and to assemble the mathematical programming model. Subsequently, the algorithm solves the generated MI(N)LP problem. Next, the algorithm compares the objective function values of the current and, if available, the previous solution: If the current solution is better than the previous one, the superstructure is expanded to incorporate more redundant units, and another optimization is performed; if not, the loop is terminated.

The algorithm can be initialized with a user-defined minimum number of units; otherwise, it starts with a superstructure incorporating one unit of each plausible technology. The employed expansion strategy increases the number of units by one in each loop. More precisely, one unit is added for each type of technology within each topographically defined site if all available units are used in the optimal solution. This process is repeated until one spare unit of each technology is available in the superstructure. The successive superstructure expansion is performed separately for each type of technology within each topographically defined site: If, e.g., a topographic constraint enforces the consideration of two decentralized cooling systems, the proposed approach increases the number of chillers for both cooling systems independently of each other. Thus, the final superstructure incorporates exactly one more unit of each technology in each topographically defined site than embedded in the optimal solution. In the author's experience, this simple strategy yields the global optimal solution: In fact, in all test cases, excessive superstructure expansion beyond the described termination criterion did not result in any better solutions.

4.3 Illustrative grassroots example

As illustrative example, the same test problem is considered as in section 3.3. The objective function is the net present value (NPV). The considered site comprises one heating and one cooling demand (Fig. 4.8). The site is connected to the regional natural gas grid (gas tariff: 6 ct/kWh) and the regional electricity grid (electricity tariff: 16 ct/kWh; feed-in tariff: 10 ct/kWh). For net present value calculations, a cash flow time of 10 years and a discount rate of 8 % are assumed.

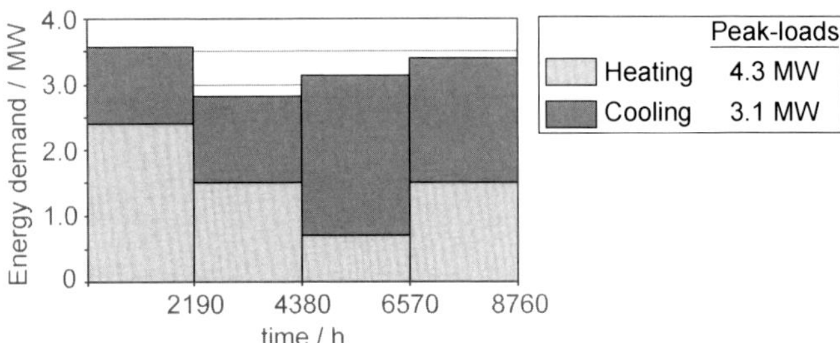

Figure 4.8: Seasonal-averaged demand profiles for heating and cooling of the synthesis test problem (stacked bar chart). Summer and winter peak loads are given in legend.

The problem is solved using the presented framework for automated superstructure-based DESS synthesis. The component set and the MILP formulation introduced in chapter 3 are used. The MILP problem is implemented in the modeling language GAMS version 23.7.3. It is solved to global optimality (maximal relative optimality gap: 0 %) using IBM ILOG CPLEX version 12.2. The computations are performed on an Intel Xeon X5650 2.67 GHz with 2 GB RAM and four kernels. The operation system is Windows XP (32-bit). Multi-threading is activated for CPLEX to perform parallel problem solving on all four kernels.

4.3.1 Optimal solution

The optimal solution equals the one found in section 3.3 (Table 4.1). The flowsheet of the optimal solution and the corresponding superstructure are shown in Fig. 4.9 a). The connectivity matrix of this solution is shown in Fig. 4.9 b). For clarity, technologies not selected in the optimal solution are removed from the reduced matrix. Total investments amount to 1.34 M€. Annual energy and maintenance cost add up to 0.88 M€. Therewith, the net present value amounts to −7.24 M€. A detailed list of the equipment installed in the optimal grassroots solution including nominal thermal powers, investment costs, operating times, and annual average part-loads can be found in appendix C.2. Excessive superstructure expansions up to five units of each technology do not result in any better solutions, and thus the discussed solution is assumed to be the global optimal solution.

Table 4.1: Optimal solution of the illustrative grassroots synthesis problem. The objective function is the net present value (NPV).

NPV	Solution structure (equipment sizing in kW)							
/ M€	B1	B2	CHP1	CHP2	TC1	TC2	AC1	AC2
−7.24	1900	100	2300	0	1900	840	370	0

In optimal configuration, the heating system consists of two boilers and one CHP engine. For cooling, two compression chillers and one absorption chiller are installed. The CHP engine is operated year-round to meet requirements for heating, and to supply driving heat for the absorption chiller during summer. Moreover, a considerable amount of the produced electricity is used on-site to drive the compression chillers. The two boilers are reserved to solely supply heating during winter. Turbo-chiller 1 is operated year-round, while turbo-chiller 2 is operated only during spring and summer. The absorption chiller is reserved to solely meet cooling requirements during summer. In Fig. 4.10, the annual average operating points of the installed equipment are shown: The CHP engine is on average operated at 70 % part-load to enable year-round operation. The peak-load absorption chiller is on average operated at 94 % part-load.

Installation of redundant units allows for load sharing enabling to run both boilers and both turbo-chillers close to their optimal operating points year-round.

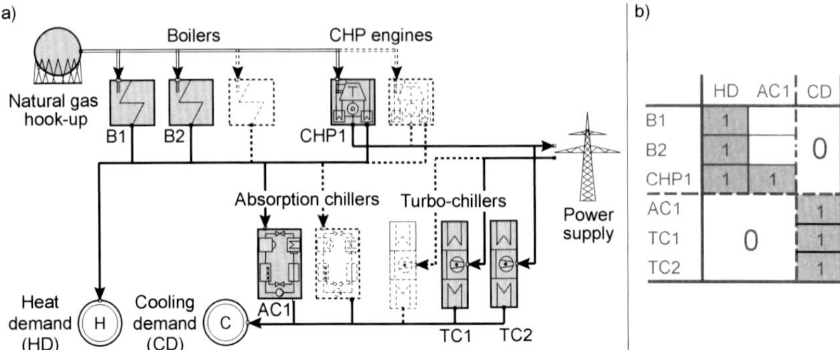

Figure 4.9: a) Final superstructure and optimal solution (gray units) for the illustrative grassroots synthesis problem. b) Reduced connectivity matrix (transposed) representing the optimal grassroots solution. For simplicity, electricity and natural gas are not shown.

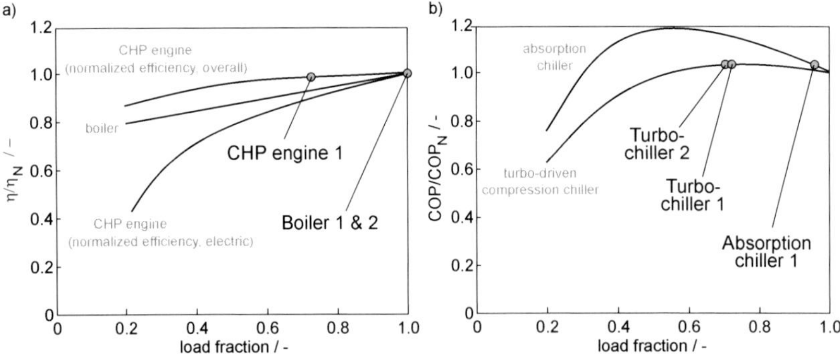

Figure 4.10: Annual average operating efficiencies (η/η_N, COP/COP_N) against operating part-loads of equipment installed in the optimal grassroots solution.

4.3.2 Successive optimization

The progress of the successive optimization is shown in Fig. 4.11. The solution obtained in optimization run 1 (one unit of each technology in superstructure), installs all units available in the superstructure (NPV = −7.245 M€). When redundant units are considered (run 2, two units of each technology in superstructure), the optimal solution incorporates two boilers, one CHP engine, two compression chillers, and one absorption chiller (NPV = −7.24 M€). However, at this point, the solution cannot yet be identified as optimal solution. Only after a third boiler and compression chiller are added to the superstructure (run 3), does optimization not improve the solution, and one spare unit of each technology is available in the superstructure. Thus, the solution of run 3 is identified as optimal solution, and the successive approach is terminated.

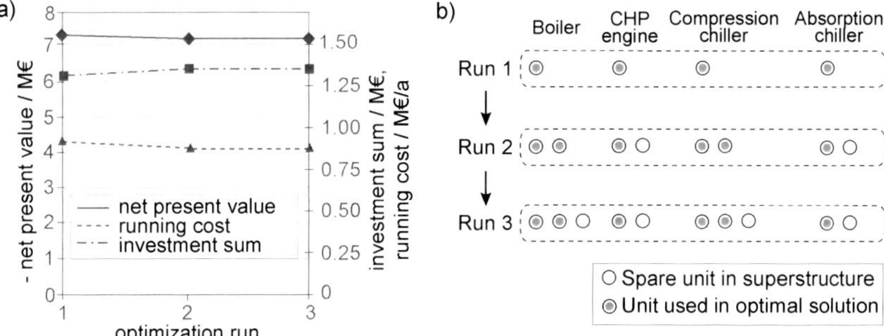

Figure 4.11: Progress of the successive optimization for the illustrative grassroots synthesis problem. a) Net present value, investment sum, and annual running cost (energy + maintenance cost) of each optimal solution plotted against the number of optimization runs. b) Schematic illustrations of number of units available in each superstructure and used in optimal configuration of each optimization run.

In this simple example, the two solutions obtained during successive optimization only marginally differ with respect to the net present value (0.14 %), which shows that the proposed methodology is capable of identifying the optimal solution as well as promising near-optimal solutions. Thus, the successive approach enables to conveniently assess the trade-offs between cost and number of units. For the simple test problem discussed in this section, the use of redundant units leads to slightly improved solutions only. However, as shown in chapter 6, for complex real-world problems, the use of redundant units leads to significant improvements.

4.3.3 Computational performance

To evaluate the computational performance of the successive optimization approach, it is compared to the so-called *naive approach*. In the naive approach, full superstructures are generated and optimized, i.e., the number of redundant units incorporated in the superstructure equals the optimization run number: in optimization run 1, one unit of each technology is incorporated in the superstructure; in run 2, two units of each technology are incorporated, etc.

The final superstructure model of the successive approach encodes 480 possible solution structures (cf. section 3.2.2):

$$\underbrace{\binom{2+3-1}{3}}_{\text{boilers}} \cdot \underbrace{\binom{4+2-1}{2}}_{\text{CHP engines}} \cdot \underbrace{\binom{2+3-1}{3}}_{\text{turbo-chillers}} \cdot \underbrace{\binom{2+2-1}{2}}_{\text{absorption chillers}} = 4 \cdot 10 \cdot 4 \cdot 3 = 480.$$

In contrast, the final superstructure of the naive approach incorporates 3 units of each technology, and thus encodes 1280 possible solution structures:

$$\underbrace{\binom{2+3-1}{3}}_{\text{boilers}} \cdot \underbrace{\binom{4+3-1}{3}}_{\text{CHP engines}} \cdot \underbrace{\binom{2+3-1}{3}}_{\text{turbo-chillers}} \cdot \underbrace{\binom{2+3-1}{3}}_{\text{absorption chillers}} = 4 \cdot 20 \cdot 4 \cdot 4 = 1280.$$

Keep in mind that these numbers do not reflect the additional decisions for equipment sizing and operation, which substantially increase the complexity of the complete synthesis problem (cf. section 3.3.3).

The total number of iterations and computation time required to identify the optimal solution using the successive approach amount to 423 000 iterations and 150 seconds (Table 4.2). A graphical illustration of the computational effort for solving the grassroots synthesis problem employing superstructures of different sizes is provided in Fig. 4.12. The number of iterations for the naive approach is plotted against the optimization run number (Fig. 4.12 a). The plot illustrates the exponential behavior of the computational effort for solving optimal synthesis problems with a growing number of considered units. In Fig. 4.12 b), the number of iterations are plotted for both the naive and the successive approach. For clarity, the number of iterations are plotted only for the first four optimization runs. Using the naive approach, the optimal solution is found in run 2. However, the solution is not identified as optimal solution before a third unit of each technology is added to the superstructure (run 3).

Table 4.2: Computational effort to identify the successive optimization solutions. Listed are the NPV, the relative optimality gap, the required number of iterations and solution time, and the structures of each solution given in numbers of installed equipment (B: boiler, CHP: CHP engine, TC: turbo-chiller, AC: absorption chiller).

run	NPV / M€	relative optimality gap	computational effort: number of iterations	computational effort: solution time /s	solution structure: B	solution structure: CHP	solution structure: TC	solution structure: AC
1	−7.25	0.14 %	5000	1	1	1	1	1
2	−7.24	0	96 000	32	2	1	2	1
3	−7.24	0	322 000	117	2	1	2	1
$\sum$	−7.24	0	423 000	150	2	1	2	1

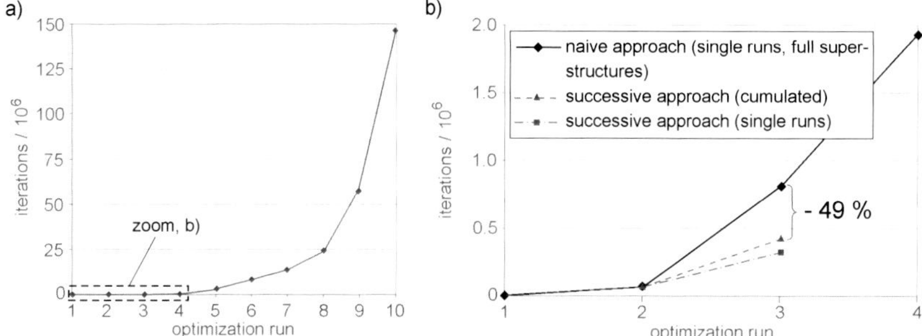

Figure 4.12: Computational effort required for solving the illustrative grassroots synthesis problem. a) Number of iterations against the number of optimization runs for the naive approach. b) Number of iterations against the number of optimization runs for the naive and the successive approach.

Employing the successive approach, the optimal solution is found in optimization run 2 as well, but again, the successive approach is terminated not until run 3 has been performed. Comparison of both approaches shows that the number of iterations required for solving the optimization problems take considerably smaller values for the successive approach: For the considered problem, the successive approach yields the optimal solution with 49 % less iterations than the naive approach.

The differences in the computational effort for solving the optimization problems of the naive and the successive approach can be attributed to the different combinatorial complexities of the problems. The discussion shows that, for larger synthesis problems, a substantial increase of the combinatorial complexity is to be expected. Thus, the strength of the successive optimization approach will grow with the problem size; or, simply put, the larger the synthesis problem, the more evident the advantage of the successive over the naive approach. Moreover, it should be emphasized that no warm starting strategy has been employed, which have the potential to significantly increase the computational efficiency of the successive approach (Ralphs and Güzelsoy, 2006) (cf. section 7.1).

4.3.4 Generation of near-optimal solution alternatives

For the considered synthesis problem, the successive approach generates two solutions, among which the decision maker can choose. The marginal difference between the two solutions' objective function values indicates a rich near-optimal solution space, which should be explored more thoroughly. For this purpose, algorithmic methods have been proposed that sequentially solve a series of optimization problems with gradually moving bounds on the objective function value (Greistorfer et al., 2008). Other approaches modify the standard branch-and-bound algorithm to directly explore the search tree (Friedler et al., 1996; Danna et al., 2007). The modified branch-and-bound algorithms efficiently generate a ranked set of integer solutions. Applied to the MILP synthesis problems considered in this thesis, these algorithms generate a vast number of solutions that differ not only with respect to equipment configuration, but also with respect to equipment sizing and operation. However, the design engineer is primarily interested in structurally different solutions, i.e., different combinations of equipment, or more technically, solutions with different values for the binary decision variables y_n representing the (non-)existence of a piece of equipment n.

To systematically generate a ranked set of integer solutions that differ with respect to equipment configuration only, so-called *integer-cut* constraints can be used (Balas and Jeroslow, 1972). To generate structurally different process solutions of process synthesis problems, integer-cut constraints were first used by Raman and Grossmann (1991). For the MILP formulation presented in this work, the following integer-cut formulation is introduced

Integer-cut based generation of structurally different solution alternatives

Assume that $(k-1)$ solutions have already been generated. Let i denote any of these solutions, then $y_n^{(i)}$ denotes the binary decision variables representing the (non-)existence of equipment n of the already known i-th best solution. For any i-th best solution, the equipment embedded in the corresponding superstructure is divided into two sets $N_1^{(i)}$ and $N_0^{(i)}$, where $N_1^{(i)}$ contains all equipment that is installed in the i-th best solution ($y_n^{(i)} = 1$), and $N_0^{(i)}$ contains all of the remainder equipment, i.e., the spare units available in the superstructure:

$$N_1^{(i)} = \{n : y_n^{(i)} = 1\},\ N_0^{(i)} = \{n : y_n^{(i)} = 0\}.$$

Based on the $(k-1)$ already generated solutions, the binary decision variables $y_n^{(k)}$ of the next k-th best solution must fulfill the following constraints:

$$\sum_{n \in N_1^{(i)}} \left(\overset{1}{\cancel{y_n^{(i)}}} - y_n^{(k)} \right) + \sum_{n \in N_0^{(i)}} \left(\overset{0}{\cancel{y_n^{(i)}}} + y_n^{(k)} \right) \geq 1 \quad \forall\, i = 1, \ldots, k-1. \tag{4.1}$$

If these constraints are added to the problem, already identified solutions become infeasible, hence forcing optimization to identify the next best, i.e., the k-th best, solution.

To understand the basic concept of this formulation, assume that the next generated k-th solution is identical to the already known solution i,

$$y_n^{(i)} = y_n^{(k)}, \quad \forall\, n.$$

In that case, both summations of Eq. (4.1) become zero, and thus the constraint is violated, i.e., the k-th solution is infeasible. Each change in the solution structure will lead to feasible solutions: If one piece of equipment is removed from the k-th solution, the first summation is incremented by one; likewise, if one new piece of equipment is installed in the k-th solution, the second summation is incremented by one.

An equivalent, however more compact, formulation for the proposed integer-cut constraints has recently been presented by Fazlollahi et al. (2012):

$$\sum_{n=1}^{n_{\max}} y_n^{(k)} \cdot \left(2\, y_n^{(i)} - 1\right) \leq \left(\sum_{n=1}^{n_{\max}} y_n^{(i)} \right) - 1 \quad \forall\, i = 1, \ldots, k-1. \tag{4.2}$$

This formulation avoids the case differentiation in the implementation of Eq. (4.1), and can therefore be implemented more efficiently.

In this work, the integer-cut constraints (4.2) are employed sequentially to automatically generate structurally different, near-optimal solutions: Starting with the optimal solution identified by the successive approach, a series of optimization problems is solved, each extended by an integer-cut constraint excluding the already known solution structures from consideration. The user can either specify the number of solutions to be generated, or the user sets a bound on the maximal relative optimality gap by which the objective function values of the generated solutions are allowed to differ from the optimal objective function value. The first input is directly translated into the number of optimization runs to be performed; the latter requires to solve a possibly large series of optimization problems until the last generated solution violates the bound on the objective function value, and thus the integer-cut approach is terminated.

To limit computational effort, the integer-cut approach is not used together with the successive superstructure expansion strategy. Instead, it is applied to the final superstructure model of the successive approach, thus limiting the identified solutions to those embedded in the final superstructure. Keep in mind that this superstructure incorporates one spare unit of each technology within each topographically-defined site; for this reason, the author expects that limiting the generation of solution alternatives to the final superstructure not to be too restrictive for the purpose of generating structurally different, near-optimal solution alternatives. In fact, as shown for the real-world problem (cf. section 6.3), this approach generates many near-optimal solution alternatives for synthesis problems of practical size.

Exemplary application of the integer-cut approach

In the following, the integer-cut approach is used to generate structurally different, near-optimal solutions for the illustrative synthesis problem. For illustration, the integer-cut approach is used to generate all solution alternatives with objective function values that differ less than 1 % from the optimal objective function value (Table 4.3). For this purpose, starting with the solutions identified by the successive approach, nine more optimization runs are performed employing the more and more constrained model of the final superstructure. Almost 2 % of the solution structures embedded in the final superstructure (9 out of 480, cf. section 4.3.3) represent near-optimal solution alternatives.

In summary, it is shown that the integer-cut approach enables to automatically and systematically generate structurally different, near-optimal solution alternatives. However, the computational effort to generate these solutions is significant: Compared

Table 4.3: Near-optimal solutions of the illustrative grassroots synthesis problem. Objective function value (NPV), relative optimality gap, computational effort, and solution structure. Solution structures given in number of installed equipment (B: boiler, CHP: CHP engine, TC: turbo-chiller, AC: absorption chiller).

k-th best solution	NPV / M€	relative optimality gap	computational effort: iterations / 10^3	computational effort: solution time /s	solution structure: B	solution structure: CHP	solution structure: TC	solution structure: AC
1st	−7.24	0	420*	150*	2	1	2	1
2nd	−7.236	0.01 %	280	106	2	1	1	1
3rd	−7.245	0.14 %	– **	–	1	1	1	1
4th	−7.250	0.21 %	300	128	1	1	2	1
5th	−7.257	0.30 %	460	211	2	1	3	1
6th	−7.270	0.48 %	380	192	3	1	2	1
7th	−7.274	0.54 %	390	151	1	2	2	0
8th	−7.275	0.55 %	340	139	2	2	2	0
9th	−7.281	0.64 %	520	252	1	2	3	0
10th	−7.292	0.79 %	300	134	3	1	3	1
11th	−7.310	1.04 %	250	104	3	2	2	0

* Best solution: computational effort of the successive approach.

** Intermediate solution of the successive approach (no additional computation).

to the successive optimization, the computational effort required to identify the nine near-optimal solutions is increased by factor 8 in terms of iterations (3 390 000 versus 423 000 iterations) and by factor 10 in terms of computation time (1570 versus 150 seconds). In other words, the generation of each next best solution requires roughly the same number of iterations and computation time as the identification of the optimal solution. Principally, this is to be expected because every integer-cut constraint excludes only a single structure from consideration, thus reducing the combinatorial complexity of the synthesis problem only marginally. Hence, in particular for large-scale and real world problems, the generation of near-optimal solution alternatives will be computationally involved as shown in chapter 6.

Analysis of near-optimal solution set

The optimal solution represents a trigeneration system that maximizes the net present value. If not for some practical other reason, the decision maker should implement this solution. However, the knowledge of the generated near-optimal solution alternatives enables the decision maker to account for further aspects that either have been neglected in the mathematical model, or that might change in the future. In the following, three near-optimal solution alternatives are analyzed with respect to such considerations:

- The 3rd best solution is an interesting alternative due to the reduced complexity of equipment installation and control. It stands out because it does not incorporate any redundant units. Thus, this solution saves capital cost at the expense of slightly increased energy cost.
- In the 7th best solution, no absorption chillers are installed. This solution might be an interesting alternative, if the installation of different chiller technologies should be avoided, or if the heating network must be expanded to integrate absorption chillers.
- In contrast to the 3rd best solution, the 10th best solution incorporates especially many units, thus saving energy cost at the expense of increased capital cost. This solution opens up greater flexibility with regard to decentralization options, equipment operation, or future system up- and downscaling.

4.4 Summary and conclusions

In this chapter, a framework is proposed for the automated superstructure-based synthesis of distributed energy supply systems. The framework employs a superstructure-generation algorithm to automatically incorporate multiple redundant units and topographic constraints into the superstructure. A successive optimization approach continuously expands and optimizes superstructures to incorporate additional units until no further improvements are observed, and thus the optimal solution is identified. The proposed algorithm employs a generic component-based modeling framework to automatically derive the mathematical programming model representing the generated superstructures. In the present implementation, the MILP formulation presented in chapter 3 is used.

The proposed methodology is successfully applied to a simple grassroots synthesis problem. The presented approach automatically evaluates the complex trade-offs inherent to DESS synthesis problems to identify the optimal solution. To enable computationally efficient problem solving, the number of units embedded in the superstructure is limited as much as possible. The successive optimization algorithm is compared to a naive approach. Both algorithms identify the same optimal solution, but the successive approach requires 49 % less iterations. For the considered case study, the successive approach yields one near-optimal solution alternative besides the optimal solution. To systematically generate a ranked set of further near-optimal candidate solutions, the integer-cut approach is proposed. In the considered case study, the ten best solutions lie within an optimality gap of 1 % with regard to the net present value.

In summary, the presented approach fulfills the requirements for optimization-based synthesis of distributed energy supply systems as discussed in section 2.5 (Table 4.4). It avoids both the *a priori* definition of a superstructure and the manual definition of many technology-specific replacement rules. In chapter 6, the proposed framework is used to solve a real-world synthesis problem presenting further features, such as retrofit synthesis, consideration of constructional limitations, and multi-objective decision support.

Table 4.4: Comparison of the requirements for an automated synthesis method as discussed in section 2.5 and the features presented in this chapter.

Requirement (cf. section 2.5)	**Feature** (☐/☒)	
• Generic automated synthesis methodology	☒	Successive approach for automated superstructure generation and expansion.
• Accounting for characteristics of distributed energy supply systems	☒	Algorithm for automated superstructure and model generation, employed MILP formulation (cf. chapter 3).
• Near-optimal solutions	☒	Integer-cut approach.
• Multi-objective optimization	☐	See chapter 6.
• Real-world synthesis	☐	See chapter 6.
• Comparison deterministic/ metaheuristic optimization	☐	See chapter 6.

Chapter 5

Superstructure-free synthesis of distributed energy supply systems

A superstructure-free optimization approach is proposed for optimal synthesis of distributed energy supply systems (DESS). This approach employs a hybrid optimization algorithm that combines metaheuristic optimization with deterministic optimization for simultaneous alternatives generation (synthesis level) and optimization (design and operation level, cf. section 2.1.2). For metaheuristic optimization, an evolutionary algorithm is used based on a mutation operator that randomly replaces components from a candidate solution by alternative designs. To avoid the manual definition of a multitude of technology-specific replacement rules, a generalized, knowledge-integrated approach is developed. For this purpose, a hierarchically-structured graph is designed, the so-called *energy conversion hierarchy* (ECH). The ECH classifies energy conversion technologies according to their functions. This allows for an efficient definition of all reasonable connections between the regarded technologies. A minimal set of generic replacement rules is then sufficient to efficiently generate all candidate solutions by structural mutations. Moreover, the set of technologies contained in the hierarchy can easily be extended. The presented approach fulfills the requirements for optimal synthesis of distributed energy supply systems as discussed in section 2.5. The proposed mutation operator is based on earlier publication by Voll et al. (2012).

First, the hierarchy-supported mutation operator is introduced (section 5.1). Next, the hybrid optimization approach is presented, which embeds the proposed mutation operator (section 5.2). Afterwards, a model-based implementation of the mutation operator as a graph grammar is proposed (section 5.3). Thereafter, the superstructure-free optimization method is applied to a simple synthesis problem (section 5.4). A brief summary concludes this chapter (section 5.5).

5.1 A hierarchy-supported mutation operator

The proposed mutation operator employs a generalized, knowledge-integrated approach to randomly replace components from a candidate solution by alternative designs. For this purpose, a hierarchically-structured graph is designed, the so-called *energy conversion hierarchy* (ECH) (section 5.1.1). The ECH enables the application of generic replacement rules to mutate flowsheet representations of distributed energy supply systems (section 5.1.2). For this purpose, the ECH classifies energy conversion technologies according to their functions. For structural mutation, generic replacement rules can then be applied to a set of technologies sharing certain functions rather than defining a multitude of technology-specific replacement rules.

5.1.1 Energy conversion hierarchy (ECH)

The energy conversion hierarchy (ECH) is divided into three levels: the meta level, the function level, and the technology level (Fig. 5.1). Nodes arranged on the *meta level* (meta nodes) represent replacement rules. Nodes arranged on the *technology level* (technology nodes) represent energy conversion technologies. Nodes arranged on the *function level* (function nodes) allow to classify energy conversion technologies according to their main functions (solid line) and types of drive (dashed line); furthermore, the function nodes establish connections to the meta nodes to define which replacement rules are applicable.

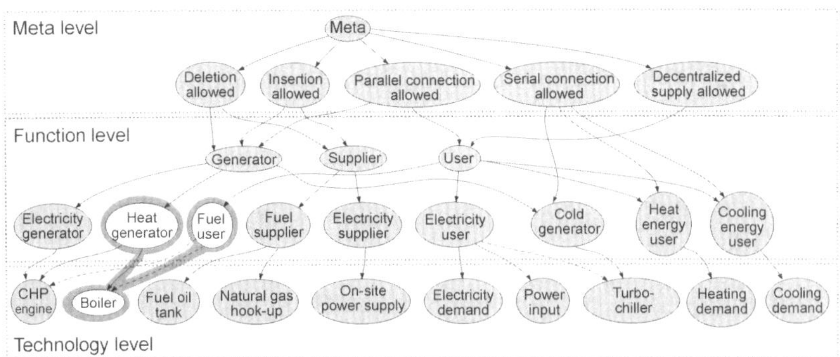

Figure 5.1: Energy conversion hierarchy.

For the classification of the energy conversion technologies, an inheritance-relation is established between the corresponding technology and function nodes: As an example, a boiler is derived from the nodes "Heat generator" (main function) and "Fuel user" (type of drive), (see Fig. 5.1, highlighted nodes). To define which replacement rules are applicable for each technology, the function nodes are linked to the corresponding meta nodes: For instance, the node "Heat generator" is linked to the meta node "Parallel connection allowed", but not to the meta node "Serial connection allowed". Thus, a boiler can be connected to another boiler in parallel with respect to its heating water cycle connections, but two boilers must not be connected to each other in serial. Through this mechanism, the generation of meaningless design alternatives is avoided. For technologies like chillers, for which it might be reasonable to set up serial connections, the corresponding links to the meta level enable this kind of connection. Note that the user is able to deactivate certain connections within the hierarchy. This might be useful, e.g., if the only fuel available on an industrial site is natural gas, and thus it will be unreasonable to ever delete the natural gas hook-up from the flowsheet. In this case, the user only needs to tag this particular component as *not deletable*.

For clarity, another type of hierarchy relation, the so-called *inverse-relation*, is not illustrated in Fig. 5.1. The information represented by this inverse-relation is used to identify "Generators" and "Suppliers" for the fulfillment of energy demands induced by "Users": As an example, if any kind of "Fuel user" is added to the flowsheet, it has to be connected to a "Fuel supplier" to generate a valid flowsheet. The specific "Fuel supplier" is chosen randomly (cf. section 5.1.4).

It should be noted that the ECH facilitates the extension of the component set considered for optimization. To integrate a new technology, the designer only needs to place a new node on the technology level and link it to appropriate function nodes. In particular, no technology-specific rules have to be specified.

5.1.2 Generic replacement rules

The information represented by the energy conversion hierarchy is used to apply generic replacement rules to particular energy conversion technologies contained in a candidate solution to generate alternative solution structures by structural mutations. The use of the ECH limits the number of required replacement rules to a set of six simple, however meaningful, replacement rules that enable to synthesize any reasonable energy supply system:

1. Remove one component with all of its interconnections.
2. Remove one component and short-circuit all of its interconnections.
3. Delete one component and insert another component.
4. Delete one component and insert a parallel connection of two other components.
5. Delete one component and insert a serial connection of two other components.
6. Delete one component and insert a component driven by a decentralized energy conversion unit.

While a formal proof of completeness of the presented rules set is beyond the scope of this thesis, it should be pointed out that, for all synthesis problems considered in this thesis, the superstructure-free synthesis methodology is capable of identifying the optimal and many near-optimal solutions.

5.1.3 Post-processing for completion of mutation

A post-processing algorithm employs the energy conversion hierarchy to inspect the created design alternatives and, if necessary, to establish missing connections and insert further components (Fig. 5.2).

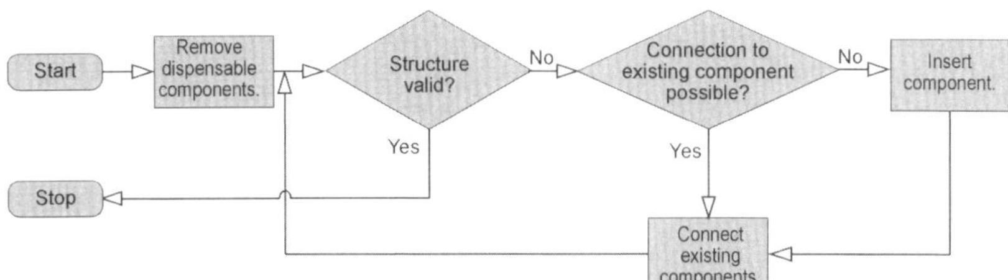

Figure 5.2: Flow diagram of the post-processing algorithm.

The algorithm starts by removing any components that are present in the flowsheet without serving their main function: As an example, a boiler powering an absorption chiller becomes dispensable when the absorption chiller is replaced by a compression chiller. Next, the algorithm checks if any unconnected components exist. If so, the algorithm checks for each of these components how to establish valid connections. For this purpose, the algorithm first tries to establish connections to any of the components already existing on the current flowsheet. If this is not possible, the algorithm inserts a new component to close the open connections. To avoid the generation of more unconnected links, the algorithm selects the component to be inserted by its main function; furthermore, components are preferred that have only this one main function. If no suitable component exists, a randomly selected component is inserted. The post-processing algorithm is continued until a valid design alternative has been created.

Moreover, the post-processing considers topographic constraints in the synthesis problem as necessary to model constructional limitations. If, e.g., a topographic constraint is defined that enforces decentralized cooling supplies, i.e., that does not allow one and the same chiller unit to be simultaneously connected to two cooling demands, the post-processing will establish only one connection from a newly inserted chiller to one of the two cooling demands.

In the author's experience, the knowledge introduced through the ECH largely avoids meaningless design alternatives, and thus the presented post-processing algorithm is mostly only used for the trivial task of establishing connections to existing components, or for regarding topographic constraints.

5.1.4 Example mutation step

For illustration of the hierarchy-supported approach, a single mutation step is discussed for a simple heating system serving as initial flowsheet for mutation (Fig. 5.3 a). In this example, a gas-fueled boiler supplies heating water to a heating demand. Natural gas is provided by an on-site natural gas hook-up. Electricity can be drawn from and fed-in to an on-site power supply.

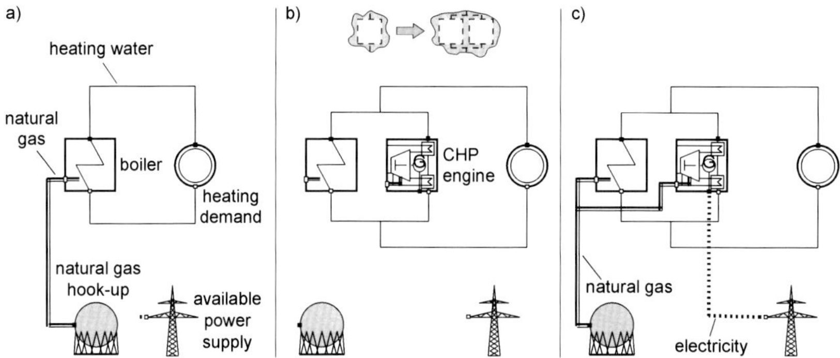

Figure 5.3: Example run of a single mutation step: a) initial flowsheet, b) flowsheet after application of replacement rule, c) flowsheet after post-processing.

The mutation operator starts with a random selection of one component to be removed from the initial flowsheet. Here, the only removable component is the boiler. Next, the algorithm consults the ECH to select one of the possible replacement rules for this component, i.e., a replacement rule that conforms the meta node connections of this component. Here, rule 4 ("delete one component and insert a parallel connection of two other components") is selected. The mutation operator now consults the energy conversion hierarchy to identify two components that may be inserted in place of the boiler. These components must fulfill two conditions: First, they must conform to the chosen replacement rule, i.e., they have to be connected to the meta node "Parallel connection allowed". Second, they must share the boiler's main function, i.e., they must be derived from the function node "Heat generator". If no component is available in the ECH that fulfills both conditions, the mutation step is restarted. In the present case, the mutation operator can choose between boilers and CHP engines. It randomly chooses one boiler and one CHP engine, and thus specifies the rule as "replace *the existing boiler* by a parallel connection of *one boiler* with *one CHP engine*". Finally, the mutation operator removes the boiler from the initial flowsheet and inserts a parallel connection of one boiler and one CHP engine. The connections corresponding to the components' main functions can be kept (Fig. 5.3 b). The missing connections (natural gas to the boiler and CHP engine, and feed-in electricity from the CHP engine) are established by the post-processing algorithm, which concludes the mutation step (Fig. 5.3 c).

5.2 A hybrid optimization approach

Hybrid optimization algorithms combine metaheuristic optimization and deterministic optimization techniques to exploit the advantages of both types (Puchinger and Raidl, 2005; Raidl, 2006). To use the hierarchy-supported mutation operator for optimal DESS synthesis, a hybrid optimization approach is implemented. For this purpose, a bi-level formulation is proposed that decomposes the DESS synthesis problem into an upper level problem that is addressed by metaheuristic optimization, and a lower level problem that is addressed by deterministic optimization.

5.2.1 Bi-level formulation

The general mathematical programming problem for single-objective DESS optimization is given by (cf. section 3.2)

$$\min_{s,d,o} f(s,d,o), \qquad s \in S,\ d \in D,\ o \in O. \tag{5.1}$$

In this formulation, the decision variable vectors s, d, and o belong to the continuous and/or integer variable spaces S, D, and O, which represent the synthesis, design, and operation decision variable spaces, respectively. As discussed in section 2.1.2, the three synthesis levels feature an inherent hierarchical structure, and thus the mathematical programming formulation can be decomposed into an upper level dealing with the synthesis, and a lower level dealing with the design and operation,

$$\underbrace{\min_{s}\left(\min_{d,o} \quad \underbrace{f(s,d,o)}_{=:f^{(s)}(d,o)}\right)}_{=:\hat{f}(s)}. \tag{5.2}$$

Thus, the mathematical programming formulation can be reformulated as

$$\begin{aligned} \min_{s} \quad & \hat{f}(s), \\ \text{s.t.} \quad & \min_{d,o} \quad f^{(s)}(d,o). \end{aligned} \tag{5.3}$$

In the following, this bi-level decomposition is applied to the MILP formulation introduced in section 3.2 (Eqs. (3.4)-(3.20)) that maximizes the net present value C_{t_CF}. The decomposed MILP formulation is given by

$$\begin{aligned} \max_{y} \quad & \hat{C}_{t_\mathrm{CF}}(y), \\ \text{s.t.} \quad & \max_{\dot{V}_\mathrm{N},\gamma,\delta_t,\dot{V}_t} \quad C^{(y)}_{t_\mathrm{CF}}(\dot{V}_\mathrm{N},\gamma,\delta_t,\dot{V}_t), \\ & y \in S, \quad \dot{V}_\mathrm{N},\gamma \in D, \quad \delta_t, \dot{V}_t \in O. \end{aligned} \tag{5.4}$$

The hybrid optimization approach (Fig. 5.4) is based on this bi-level formulation (5.4). However, the structural decisions y are not explicitly modeled in a superstructure. Instead, the presented mutation operator is embedded in an evolutionary algorithm (cf. section 2.2.2) that continuously evolves new configuration alternatives to perform optimization on the synthesis level. For equipment sizing and operation, rigorous MILP optimization is used as local refinement strategy; i.e., for each configuration alternative generated by mutation, an MILP problem is solved to identify the optimal equipment sizing and operation that maximizes the net present value. The problem formulation of the hybrid optimization is given by

$$\begin{aligned} \max_{\sigma} \quad & \hat{C}_{t_\mathrm{CF}}(\sigma), \qquad \sigma \in \Sigma, \\ \text{s.t.} \quad & \max_{\dot{V}_\mathrm{N},\gamma,\delta_t,\dot{V}_t} \quad C^{(\sigma)}_{t_\mathrm{CF}}(\dot{V}_\mathrm{N},\gamma,\delta_t,\dot{V}_t), \end{aligned} \tag{5.5}$$

where σ represents a structure evolved by mutation, and Σ represents the set of all possible structures.

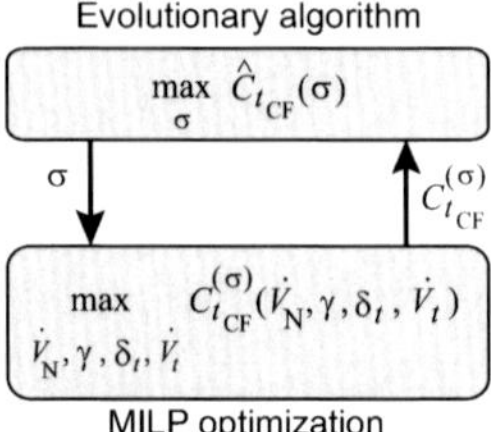

Figure 5.4: Hybrid optimization approach.

In this thesis, the hybrid optimization is based on the MILP formulation introduced in chapter 3.2. However, it should be noted again that the generic component-based modeling enables to use any other programming formulation as well.

5.2.2 A simplified mutation operator for MILP-based synthesis

The employed MILP formulation is based on simple energy balances neglecting temperature and pressure levels of the energy carriers. Therefore, effects such as mixing temperatures and temperature-dependent performance characteristics are neglected as well. Hence, different equipment interconnections, i.e., parallel and serial connections, do not affect the system performance, and can thus be neglected. In addition, transmission losses and the network layout are neglected, and thus the distinction between centralized and decentralized plant layouts makes no difference for the optimization. In turn, when employing this MILP framework for DESS synthesis, it is sufficient to use a simplified version of the hierarchy-supported mutation operator neglecting these model details.

In the simplified ECH (Fig. 5.5), the distinction of the meta nodes "Parallel connection allowed" and "Serial connection allowed" is omitted, and rather represented by the meta node "Parallel connection allowed". Moreover, the meta node "Decentralized supply allowed" is omitted as well. Accordingly, the simplified set of generic replacement rules is reduced to the following four rules:

1. Remove one component with all of its interconnections.
2. Remove one component and short-circuit all of its interconnections.
3. Delete one component and insert another component.
4. Delete one component and insert a parallel connection of two other components.

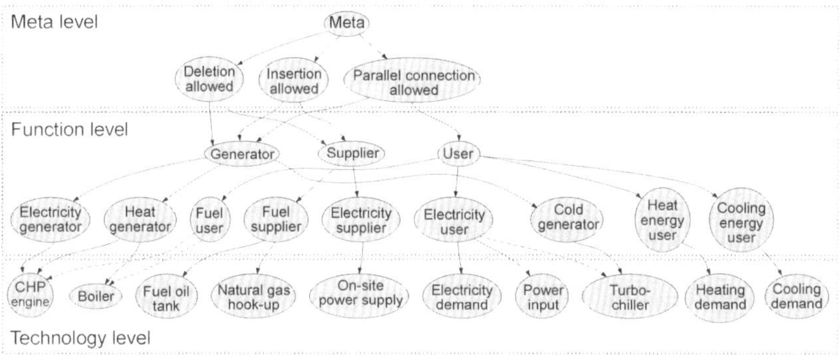

Figure 5.5: Simplified energy conversion hierarchy.

5.3 Model-based implementation of the mutation operator

To use the proposed mutation concept in an evolutionary algorithm, a suitable mutation operator needs to be implemented. In this section, a model-based implementation as *graph grammar* is presented. This implementation is particularly suited because it enables the design engineer to use the ECH as graphical modeling language to modify the component set considered for optimization. In the following, the concept of graph grammars is introduced briefly. Next, the ECH-based graph grammar design is presented. Last, general design principles for mutation operators are introduced, and the designed graph grammar is evaluated with regard to these principles. An in-depth introduction to graph grammars is given by Rozenberg (1999) with a chapter on chemical engineering applications by Cremer et al. (1999).

5.3.1 Graph grammars

Graphs can be used as intuitive models of discrete and complex systems (Bondy and Murty, 1976). For graph-based representation of energy systems models, the nodes represent energy conversion units and the edges represent the mass and energy flows between these units. Graph grammars deal with the transformation of graphs by application of so-called *productions*. A production defines the transformation of a starting graph, the so-called *host graph* (H), into a new graph. A production consists of a left-hand side, the *mother graph* (M), a right-hand side, the *daughter graph* (D)

and an *embedding mechanism* (E). Thus, a production applies E to replace M from H by D. A set of productions is called a *production set*. Productions typically specify changes in small parts of the host graph, such as adding parts to it, deleting parts from it, or exchanging parts of it. To apply a production to a host graph, the host graph must be searched for an occurrence of the mother graph. Searching the host graph for mother graphs is in general a computationally complex problem. Fortunately, for the present application, this problem can be considerably simplified by considering only mother graphs consisting of single labeled nodes, referred to as *symbols*. In graph grammars, *terminal* and *nonterminal* symbols are distinguished: Terminal symbols represent final states of a graph's nodes. Nonterminal symbols represent intermediate states of a graph's nodes enabling transformations between terminal symbols. A valid graph exclusively consists of terminal symbols. If a production is applied to a terminal symbol, a nonterminal symbol is created. The generated graph is not valid until the nonterminal symbol is transformed into a terminal symbol.

The production set of a graph grammar can be illustrated either by listing all single production rules, or in tree form: The six productions shown in Fig. 5.6 a) represent possible transformations between the terminal symbols "A" and "B", and their common nonterminal symbol "*".

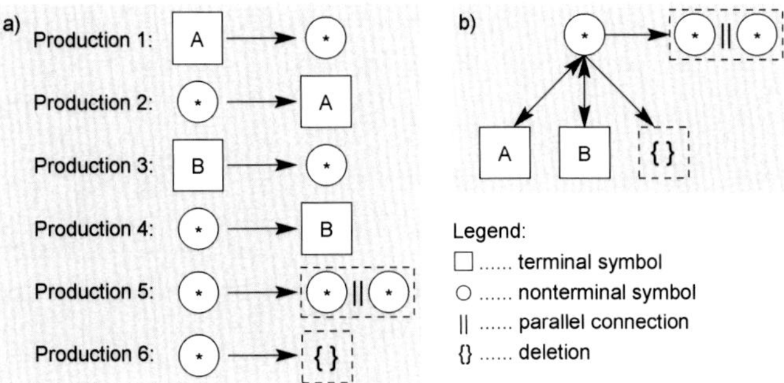

Figure 5.6: Different representations of one and the same production set comprising six productions: a) list representation, b) tree representation.

In Fig. 5.6 b), a tree is shown representing the same production set: An edge in the tree represents the existence of a production between two nodes. The arrows on these edges specify the direction of the productions, i.e., from which to which nodes they can be applied. The tree representation enables a clearer and more compact representation of the production set.

5.3.2 Model-driven graph grammar design for mutation

To model the set of replacement rules applied during mutation, a *model-driven* approach is employed for the ECH-based graph grammar design. According to Stahl and Völter (2006), *model-driven software development* (MDSD) aims at simplifying the development process while improving both software quality and reusability; therefore, MDSD increases the level of abstraction in programming, e.g., by using *domain specific languages* (DSL), *model transformations* and *code generations*. DSLs enable domain experts to build software systems by performing system modeling rather than programming. In the presented approach, the modeling language is the energy conversion hierarchy.

For illustration, the model-driven graph grammar design is presented for a reduced version of the simplified ECH (Fig. 5.7). The technology level of the reduced ECH is limited to the nodes modeling the conversion of natural gas into heat and electricity through boilers and CHP engines. First, all terminal and nonterminal symbols are created: Terminal symbols represent particular energy conversion technologies (technology nodes), and thus a terminal symbol is created for each technology node. Nonterminal symbols represent their main functions (function nodes), and thus a nonterminal symbol is created for all corresponding main function nodes.

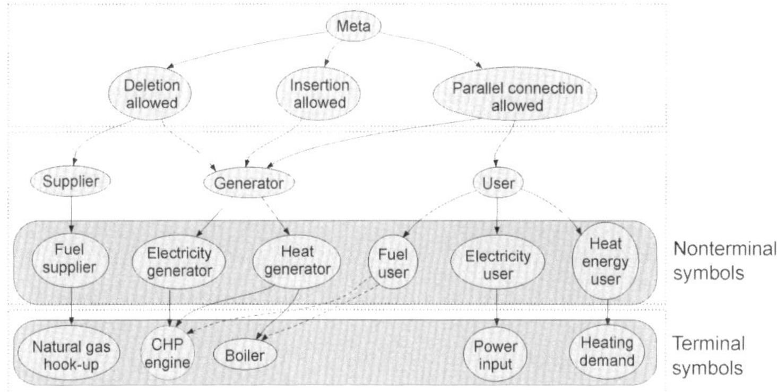

Figure 5.7: Adoption of terminal and nonterminal symbols from technology and function nodes of a reduced version of the simplified ECH (section 5.2.2).

Next, the productions are specified: The so-called *trivial productions* directly transform terminal to nonterminal symbols and vice versa; they can be directly adopted from the inheritance-relations of the corresponding technology nodes and their main function nodes: As an example, the productions transforming the terminal symbol "Natural gas hook-up" into the nonterminal symbol "Fuel supplier" and vice versa. Finally, those productions are defined allowing for deletion, insertion, and parallel connection of nonterminal symbols; these *nontrivial productions* can be derived from the links of the function nodes to the corresponding meta nodes. The resulting production set contains 17 productions (Fig. 5.8 a):

- For the nonterminal symbol "Fuel supplier", three productions are defined: two trivial productions allowing for its direct transformation to and from the "Natural gas hook-up"; and one production allowing for its deletion.
- For both nonterminal symbols "Electricity user" and "Heat energy user", only the trivial productions are specified allowing for their direct transformations to and from the nonterminal symbols "Power input" and "Heating demand", respectively. No productions are specified to generate parallel connections: This is because the ECH does not allow the insertion of these symbols.
- For the nonterminal symbols "Heat generator" and "Electricity generator", six and four productions are specified, respectively: two each allow for deletion and parallel connections; another four and two productions, respectively, allow for direct transformations to and from the terminal symbols "Boiler" (in case of the "Heat generator") and "CHP engine" (in case of both the "Heat generator" and the "Electricity generator").

To minimize the resulting production set, all nonterminal symbols are omitted that can only be transformed by trivial productions, and therefore do not contribute to structural variations. Furthermore, in the simplified ECH, the "Natural gas hook-up" is the only fuel supplier; to guarantee that, in all generated flowsheets, boilers and CHP engines will be supplied with fuel, the production allowing for the fuel supplier's deletion is also omitted. Thus, the production set is simplified to the ten productions concerned with the transformations of the nonterminal symbols "Electricity generator" and "Heat generator" (Fig. 5.8 b).

The designed graph grammar can be classified as a *node replacement graph grammar* with *node label controlled* (NLC) node replacement mechanisms; furthermore, it can be categorized as a *neighborhood-controlled embedding* (NCE) grammar: The embedding mechanisms E are limited to establish connections from daughter nodes to neighboring host nodes, i.e., host nodes to which a direct connection from the daughter node can be established via a single edge. In particular, the connections representing the ECH's main function relations can be directly adopted from the mother to the daughter graph. This considerably simplifies the embedding process (Rozenberg, 1999).

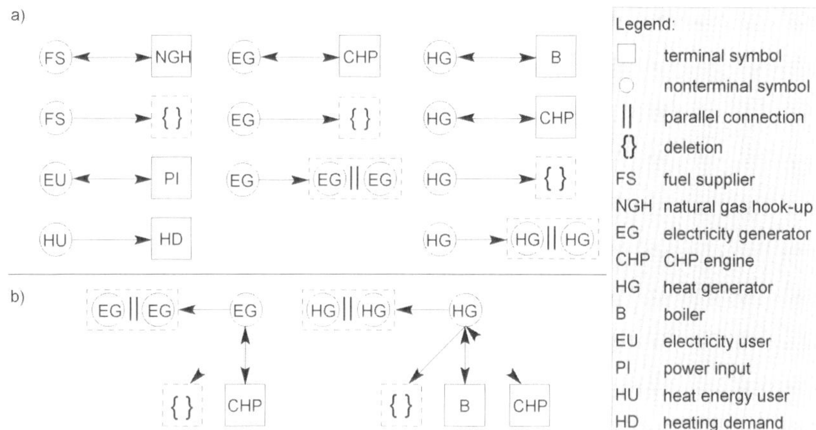

Figure 5.8: Production set derived from simplified ECH (see text for details): a) Complete production set (list representation), b) Reduced production set (tree representation).

5.3.3 Evaluation of the designed mutation operator

Usually, mutation operators are specifically designed for a class of optimization problems (Rudolph, 1994). Nevertheless, Beyer (2001) have defined general principles proven to lead to good mutation operators: These principles are *locality*, *reachability*, *scalability*, and *symmetry*. It should be noted that the importance of the listed design principles can differ depending on the specific optimization problem; also, a mutation operator violating one or more of these principles does not necessarily fail in solving a particular optimization problem.

Locality. Mutation is a variation operator used for local search. The locality principle demands that small variations should be generated with higher probabilities than large variations.

Reachability. The reachability (or ergodicity) principle guarantees the global convergence of mutation-based evolutionary algorithms. Given any point in the search space, every other point in the search space should be reachable within a finite number of mutation steps.

Scalability. If a mutation operator fulfills the reachability principle, the optimization can still get stuck in local optimal solutions. The optimization can benefit from a reachable mutation operator only if the operator also fulfills the scalability principle: Scalability requires that the strength of one mutation step should be tunable to adapt to the shape of the fitness landscape. This principle is based on the assumption that small variations in the search space result on average in small changes in the objective function values.

Symmetry. The symmetry (or unbiasedness) principle is necessary to guarantee an all-over and uniform search of the search space. While selection is used to exploit the fitness information to direct the search into promising regions, mutation is used to explore the search space. Thus, given any parental population, a mutation operator should use the parents' search space information only; it should not use any fitness information. To fulfill the symmetry principle, the mutation operator should not introduce any bias in the variation process.

In the following, the compliance of the designed mutation operator with these principles is evaluated. The following discussion is based on the representation of the mutation operator as a graph grammar.

Scalability and reachability

To satisfy the scalability condition, the number of replacements performed within one mutation step can be varied. This also leads to satisfaction of the reachability condition, because basically a very strong mutation step will replace all components from a given flowsheet and will insert a completely new flowsheet of any size.

Locality and symmetry

The locality concept demands that mutation should generate small variations with higher probabilities than large variations. The designed graph grammar is a neighborhood-controlled embedding (NCE) grammar, thus emphasizing the locality of its production set (cf. section 5.3.2). The symmetry condition demands to not introduce any bias in the search process. In the present study, the symmetry principle is met by providing an inverse counterpart to any production. Furthermore, to satisfy both the locality and symmetry conditions, the random selection of productions needs to be controlled by appropriate probabilities. For this purpose, a probabilistic graph grammar is introduced, which assigns probabilities to the single productions.

The production set in Fig. 5.9 equals the one derived in section 5.3.2 (cf. Fig. 5.8 b); only here, all productions are tagged with probabilities $P_{\mathrm{x}\to\mathrm{y}}$ defining the probability for the selection of a production replacing symbol x by symbol y.

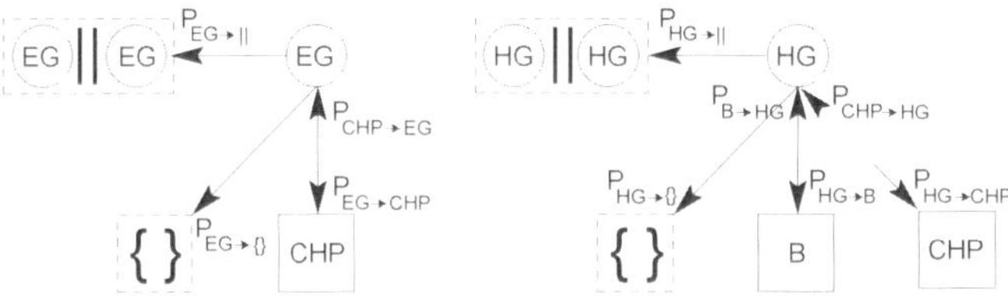

Figure 5.9: Tree representation of expanded production set with tagged production probabilities.

For each symbol, the sum of probabilities of all productions is one:

$$P_{\text{EG}\to\|} + P_{\text{EG}\to\{\}} + P_{\text{EG}\to\text{CHP}} = 1, \tag{5.6}$$

$$P_{\text{CHP}\to\text{EG}} + P_{\text{CHP}\to\text{HG}} = 1, \tag{5.7}$$

$$P_{\text{HG}\to\|} + P_{\text{HG}\to\{\}} + P_{\text{HG}\to\text{B}} + P_{\text{HG}\to\text{CHP}} = 1, \tag{5.8}$$

$$P_{\text{B}\to\text{HG}} = 1. \tag{5.9}$$

To fulfill the locality principle, the so-called *size-changing probability* $P_{\Delta\text{Size}}$ is introduced: It guarantees that productions changing the number of components in a given flowsheet are applied less likely than those simply exchanging technologies with common functions. In the illustrated production set, there are four productions changing the number of components: two, which extend a given flowsheet by establishing parallel connections (corresponding probabilities $P_{\text{EG}\to\|}$ and $P_{\text{HG}\to\|}$), and two, which reduce a given flowsheet by deleting symbols from it (corresponding probabilities $P_{\text{EG}\to\{\}}$ and $P_{\text{HG}\to\{\}}$). $P_{\Delta\text{Size}}$ allows to control how likely these productions are applied. In addition, it is assumed that size-changing productions will be applied with equal probabilities for all nonterminal symbols:

$$P_{\Delta\text{Size}} = P_{\text{EG}\to\|} + P_{\text{EG}\to\{\}} = P_{\text{HG}\to\|} + P_{\text{HG}\to\{\}}. \tag{5.10}$$

To satisfy the symmetry condition, neither the extension nor the reduction of a flowsheet should be favored, and thus the corresponding probabilities are equal:

$$P_{\text{EG}\to\|} = P_{\text{EG}\to\{\}}, \tag{5.11}$$

$$P_{\text{HG}\to\|} = P_{\text{HG}\to\{\}}. \tag{5.12}$$

Furthermore, if a terminal symbol can be transformed only into exactly one nonterminal symbol, the probability of the production equals one:

$$P_{\text{B}\to\text{HG}} = P_{\text{CHP}\to\text{EG}} = P_{\text{CHP}\to\text{HG}} = 1. \tag{5.13}$$

Last, it must be assured that the probability for the transformation of one terminal symbol to another terminal symbol equals the probability for the opposite transformation (here, transformation of the terminal symbol "Boiler" to the terminal symbol "CHP engine" via the nonterminal symbol "Heat generator", and the other way round):

$$\underbrace{P_{\mathrm{B}\to\mathrm{HG}}}_{1} \cdot P_{\mathrm{HG}\to\mathrm{CHP}} = \underbrace{P_{\mathrm{CHP}\to\mathrm{HG}}}_{1} \cdot P_{\mathrm{HG}\to\mathrm{B}}. \tag{5.14}$$

For a given $P_{\Delta\mathrm{Size}}$, the probabilities of all productions can be calculated by Eqs. (5.6) to (5.14). These probabilities guarantee the symmetry of the mutation operator; locality is guaranteed if the probabilities of the size-changing productions ($P_{\mathrm{EG}\to\|}$, $P_{\mathrm{EG}\to\{\}}$, $P_{\mathrm{HG}\to\|}$, and $P_{\mathrm{HG}\to\{\}}$) are lower than the probabilities of all other productions.

If, e.g., $P_{\Delta\mathrm{Size}} = 1/3$ is assumed, the probabilities of the productions are $P_{\mathrm{EG}\to\|} = P_{\mathrm{EG}\to\{\}} = P_{\mathrm{HG}\to\|} = P_{\mathrm{HG}\to\{\}} = 1/6$, $P_{\mathrm{EG}\to\mathrm{CHP}} = 2/3$, $P_{\mathrm{B}\to\mathrm{HG}} = P_{\mathrm{CHP}\to\mathrm{EG}} = P_{\mathrm{CHP}\to\mathrm{HG}} = 1.0$, $P_{\mathrm{HG}\to\mathrm{B}} = P_{\mathrm{HG}\to\mathrm{CHP}} = 1/3$, and thus the locality condition is satisfied.

When the ECH is expanded, the new production probabilities can be calculated automatically according to the simple calculations presented above. Thus, the generic ECH-based graph grammar design provides the basis for a component-based optimization framework and an easy expansion of the model library.

5.4 Illustrative grassroots example

As illustrative example, the same grassroots test problem is considered as in sections 3.3 and 4.3. The problem is solved using the hybrid optimization algorithm for superstructure-free DESS synthesis employing the mutation graph grammar. The objective function is the net present value. The component models introduced in chapter 3 are used. For illustration, the simplest form of $(\mu+\lambda)$-selection is used for the evolutionary algorithm with $\mu = 1$ and $\lambda = 1$; i.e., in each generation, only one offspring individual is generated by mutating the parent individual, and the better of the two individuals is kept as current best individual (parent) for the next generation. For mutation, the size-changing probability $P_{\Delta\mathrm{Size}} = 1/3$ is assumed. The mutation strength, i.e., the number of productions performed within one mutation step, is randomly chosen for each generation according to a uniform probability distribution. The minimum mutation strength is set to one. The maximum mutation strength is set to five. The evolutionary algorithm is terminated after 50 generations (including the initial solution). The parameters of the evolutionary algorithm have been chosen by trial and error rather than by elaborate algorithm tuning. As in section 4.3, the MILP subproblem for sizing and operation is solved to global optimality (maximal relative optimality gap: 0 %) using IBM ILOG CPLEX version 12.2 on a 32-bit Windows XP system. The computations are performed on an Intel Xeon X5650 2.67 GHz with 2 GB RAM and four kernels, of which all are used for parallel problem solving.

5.4.1 Synthesis with the simplified energy conversion hierarchy

First, the mutation operator uses the simplified ECH encompassing boilers, CHP engines, and turbo-driven compression chillers (cf. Fig. 5.5). The productions and their corresponding probabilities are given in appendix B.1. The provided initial solution incorporates one boiler and one compression chiller. The initial solution is not feasible, because a single boiler cannot be sized adequately to meet heating peak-loads in winter as well as minimum heat loads in summer.

The best solution found represents a cogeneration system (Table 5.1, Fig. 5.10 a). Total investments amount to 1.40 M€. Annual energy and maintenance cost add up to 0.88 M€. Therewith, the net present value amounts to −7.27 M€. A detailed list of the equipment installed in the optimal solution including nominal thermal powers, investment costs, operating times, and annual average part-loads can be found in appendix C.3.1.

Table 5.1: Optimal solution of the illustrative grassroots synthesis problem employing the simplified ECH. The objective function is the net present value (NPV).

NPV	Solution structure (sizing in kW)				
/ M€	B1	CHP1	CHP2	TC1	TC2
−7.27	1900	1500	900	1900	1210

5.4.2 Synthesis with the extended energy conversion hierarchy

In the following, absorption chillers are considered besides turbo-chillers. Thus, the ECH is extended to include absorption chillers. To do so, first an "Absorption chiller" node is added to the technology level; second, this node is linked to the function nodes representing its main function ("Cold generator") and type of drive ("Heat energy user"). This already concludes the addition of a new component to the optimization problem. The extended ECH is shown in Fig. 5.11. The benefit of the employed framework can thus already be seen from the avoidance of any technology-specific rule definitions. Due to the easy addition of technologies into the optimization problem, the suggested approach provides the basis for a component-based optimization framework using an expandable model library.

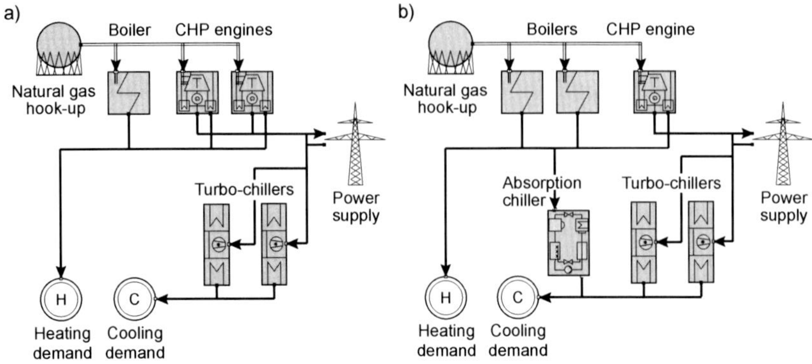

Figure 5.10: Optimal solutions of the illustrative grassroots synthesis problem employing the simplified (a) and the extended (b) energy conversion hierarchy.

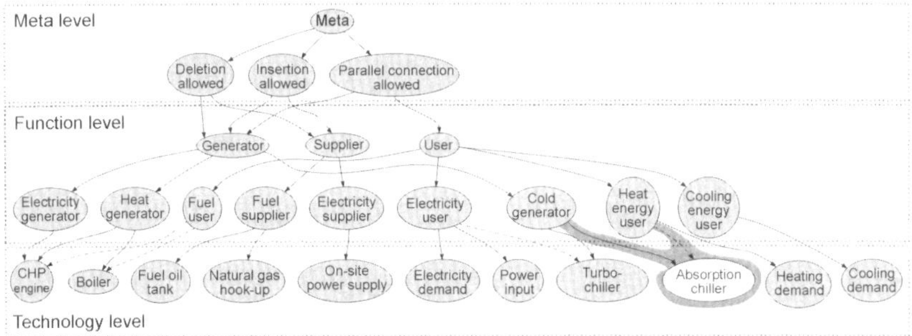

Figure 5.11: Extended energy conversion hierarchy incorporating absorption chiller.

The optimal synthesis problem is solved once more based on the extended ECH. The productions and their corresponding probabilities are given in appendix B.1. Again, the initial solution is infeasible incorporating one boiler and one compression chiller. The best solution found (Table 5.2, Fig. 5.10 b) equals the optimal solution found when the automated superstructure-based synthesis approach is used (for details, see section 4.3). The optimal solution represents a trigeneration system, in which two boilers and one CHP engine provide heating, and one absorption and two compression chillers provide cooling. The optimal solution differs from the one obtained when only compression chillers are considered. However, the net present value is only marginally improved by 0.5 % (total investments 1.34 M€, annual energy and maintenance cost 0.88 M€, and net present value −7.24 M€).

Table 5.2: Optimal solution of the illustrative grassroots synthesis problem employing the extended ECH. The objective function is the net present value (NPV).

NPV / M€	Solution structure (sizing in kW)					
	B1	B2	CHP1	TC1	TC2	AC1
−7.24	1900	100	2300	1900	840	370

5.4.3 An exemplary optimization run

In Fig. 5.12, the progress of an exemplary optimization run is shown using the extended energy conversion hierarchy. The net present value (NPV) of each offspring individual is plotted against the generation number. The infeasible initial solution is not shown in the plot. To illustrate the convergence, the current best solution is represented by the straight lines. Note that, for the employed $(1 + 1)$-selection strategy, only one individual is kept in each generation, and thus the generation number equals the individual number and also the number of function evaluations.

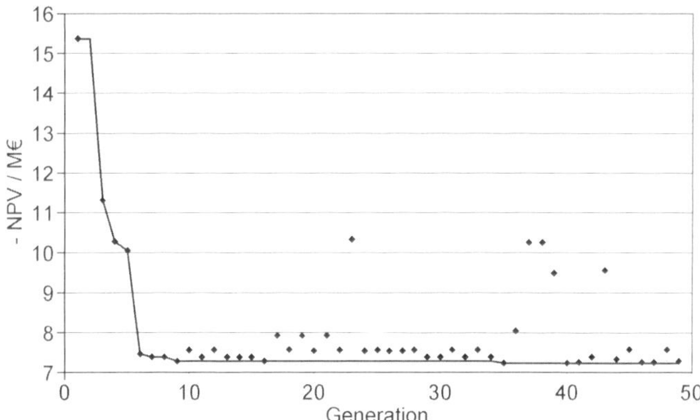

Figure 5.12: Progress of an exemplary optimization run. Net present value (NPV) plotted against generation number. The current best solution of each generation is illustrated by the straight lines.

Due to the stochastic nature of the evolutionary algorithm, not every mutation necessarily improves the current best solution. In fact, every now and then, the algorithm generates poor solutions. This is a typical characteristic of metaheuristic search algorithms and can be attributed to the structural variations caused by the randomized search. However, it should be emphasized that this behavior is rather a feature than a shortcoming because it helps to avoid the algorithm to get stuck in local optimal solutions.

In the exemplary run, an improved solution is generated in eight of the first nine generations. In the remaining 40 generations, only four more improvements are observed (16, 35, 40, and 41). Individual 41 is the optimal solution. It should be emphasized that eight of the 49 individuals generated during optimization represent structurally different, near-optimal solutions with objective function values that differ less than 1 % from the optimal objective function value (Table 5.3). Moreover, six of these eight solutions are among the ten best solutions generated by the integer-cut approach of the superstructure-based optimization method (cf. section 4.3.4). Individuals 30 and 44 stand out because they incorporate up to four redundant units per technology, and thus they represent solutions that were not generated by the integer-cut approach due to the limited superstructure.

Table 5.3: List of structurally different, near-optimal solutions of the exemplary optimization run including individual number, objective function value, relative optimality gap, and solution structure. Solution structures are given in numbers of installed equipment (B: boiler, CHP: CHP engine, TC: turbo-chiller, AC: absorption chiller).

individual	NPV / M€	relative optimality gap	solution structure			
			B	CHP	TC	AC
9	−7.275	0.55 %	2	2	2	0
16	−7.274	0.54 %	1	2	2	0
29	−7.281	0.64 %	1	2	3	0
30	−7.292	0.78 %	2	1	4	1
32	−7.292	0.79 %	3	1	3	1
35	−7.250	0.21 %	1	1	2	1
40	−7.236	0.01 %	2	1	1	1
41*	**-7.235**	**0.00 %**	**2**	**1**	**2**	**1**
44	−7.305	0.97 %	4	1	2	1

* Optimal solution.

The structural diversity of the generated near-optimal solutions supports the decision maker to take into account further aspects for the final decision that have been neglected in the mathematical model, such as complexity of the required control system, desired flexibility with regard to decentralization options, and the robustness of the diverse solutions towards variations in the energy demands and energy prices.

5.4.4 Convergence behavior

To analyze the convergence behavior of the proposed algorithm, the optimization is repeated 500 times employing the extended energy conversion hierarchy. The computation times for generating and optimizing 50 individuals vary between 60 seconds and 73 minutes! The different computation times can be attributed to the complexity of the solution structures randomly generated during optimization. The median computation time is 504 seconds (8:24 minutes). No correlation is observed between computation time and the quality of the optimal solution, i.e., optimal objective function value.

To model the typical convergence behavior, a median curve is calculated from all current best objective function values in each generation (Fig. 5.13). More reliable estimates on the convergence of the optimization can be gained from the 75th and 90th percentile curves. Similar to the median curve, the percentile curves illustrate for each generation the least best objective function values identified in 75 % and 90 % of the runs, respectively. It is shown that, in every second run (median curve), already within the first nine generations, a solution is found whose objective function value differs less than 2 % from the optimal objective function value. Moreover, the optimal objective function value is reached already within the first 20 generations. Even if the 75th percentile curve is considered, the best solution found within 50 generations has an objective function value that differs less than 0.14 % from the optimal objective function value. However, in 10 % of all runs (90th percentile curve), solutions with a relative optimality gap less than 2 % are not found until at least 37 generations have been evaluated.

Another way to illustrate the convergence behavior is to plot the frequency distribution of the current best objective function values for different generation numbers for all 500 optimization runs (Fig. 5.14). In generation 5, only very few optimization runs have already progressed towards the global optimal solution. However, already 42 % of the current best solutions have objective function values that differ less than 2 % from the optimal objective function value. In generation 20, already 61 % of all solutions lie within the near-optimal region (optimality gap smaller than 1 %). In generation 35, a distinct peak of the frequency distribution occurs for near-optimal solutions in the optimality gap range between 0 % and 0.5 %. Already 81 % of all current best solutions lie within the near-optimal solution space. Finally, in generation 50, 52 % of all runs have converged towards global optimality, and 94 % of the current best solutions have objective function values that differ less than 1 % from the optimal objective function value.

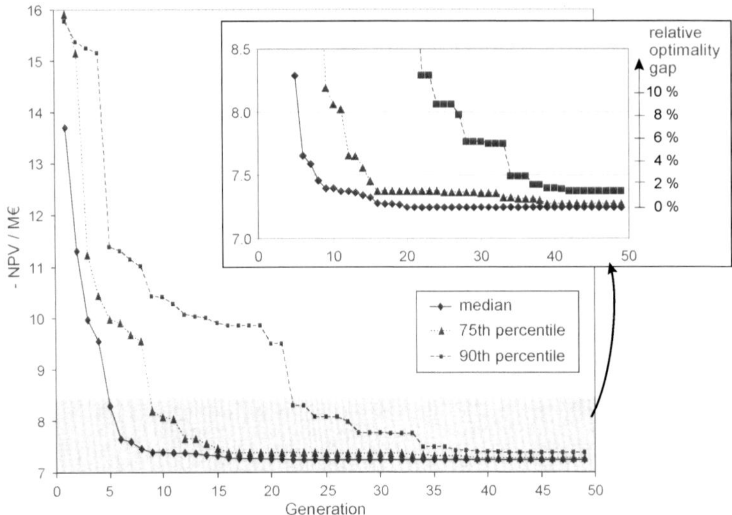

Figure 5.13: Median, 75th percentile, and 90th percentile curves of 500 optimization runs. Net present value (NPV) plotted against generation number.

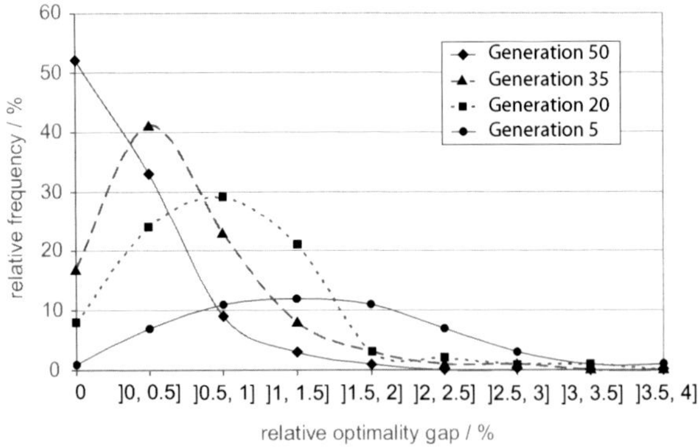

Figure 5.14: Frequency distributions of current best solutions for different generation numbers. Relative frequency plotted against relative optimality gap. Solutions with relative optimality gaps larger than 4 % are not shown.

The statistical analysis shows that the proposed superstructure-free synthesis methodology often quickly identifies near-optimal solutions. However, the evolutionary algorithm does not allow direct control of the number of near-optimal solutions generated during optimization. Moreover, global optimality cannot be guaranteed within a limited number of generations. It should be pointed out, again, that an infeasible solution is provided as initial solution. If a better initial solution is provided, enhanced convergence can be expected (Maaranen et al., 2007; Liu, 2010). Moreover, evolutionary algorithms can be tuned for particular optimization problems to exploit the problems' special characteristics: Bartz-Beielstein et al. (2005), e.g., report performance improvements of an evolutionary algorithm through algorithm tuning by more than 90 % with respect to the required number of function evaluations to identify the optimal solution.

5.5 Summary and conclusions

In this chapter, a novel approach is presented for the superstructure-free optimal synthesis of distributed energy supply systems. The proposed approach is based on a generic component-based modeling framework, which allows to use arbitrary modeling formulations. Based on the MILP formulation presented in chapter 3, a bi-level formulation is derived that decomposes the overall synthesis problem into one dealing with equipment selection (synthesis level), and another one dealing with equipment sizing and operation (design and operation level, cf. section 2.1.2). To solve this bi-level problem, a hybrid optimization approach is proposed that addresses the upper level synthesis problem by an evolutionary algorithm and the lower level design and operation problem by rigorous MILP optimization.

The evolutionary algorithm is based on a knowledge-integrated mutation operator that employs replacement rules to replace parts of energy supply systems by alternative designs. To minimize both the number of replacement rules and meaningless design alternatives generated during mutation, all relevant energy conversion technologies are classified into a hierarchically-structured graph, the so-called energy conversion hierarchy (ECH). The ECH allows for an efficient definition of all reasonable connections between the regarded components through a minimal set of generic replacement rules. Finally, all possible replacement rules are implemented in a model-driven graph grammar that applies the ECH as graphical modeling language. This facilitates the extension of the mutation operator to include further technologies in the optimization.

The proposed method is successfully applied to a simple grassroots synthesis problem. The proposed approach automatically identifies complex solutions such as cogeneration and trigeneration without *a priori* specification of the synthesis alternatives, thus demonstrating the power of the suggested optimization framework. Moreover, the proposed approach identifies near-optimal solutions without any additional computations. However, global optimality cannot be guaranteed.

In summary, the presented approach fulfills the requirements for optimal synthesis of distributed energy supply systems as discussed in section 2.5 (Table 5.4). It avoids both the *a priori* definition of a superstructure and the manual definition of many technology-specific replacement rules. In chapter 6, the superstructure-free approach is applied to a real-world synthesis problem, thus presenting further features.

Table 5.4: Comparison of the requirements for an automated synthesis method as discussed in section 2.5 and the features of the synthesis methods proposed in this thesis.

Requirement (cf. section 2.5)		**Feature (☐/☒), chapter 4 (superstructure-based)**		**Feature (☐/☒), chapter 5 (superstructure-free)**
• Generic automated synthesis methodology	☒	Successive approach for automated superstructure generation and expansion.	☒	Knowledge-integrated hybrid algorithm, energy conversion hierarchy (ECH).
• Accounting for characteristics of DESS	☒	Algorithm for automated superstructure and model generation, employed MILP formulation.	☒	Energy conversion hierarchy (ECH), employed MILP formulation (cf. chapter 3).
• Near-optimal solutions	☒	Integer-cut approach.	☒	Evolutionary algorithm, no additional computations.
• Multi-objective optimization	☐	See chapter 6.	☐	See chapter 6.
• Real-world synthesis	☐	See chapter 6.	☐	See chapter 6.
• Comparison deterministic/ metaheuristic optimization	☐	See chapter 6.	☐	See chapter 6.

Chapter 6

Real-world example

In this thesis, two methodologies are proposed for automated optimization-based synthesis of distributed energy supply systems (DESS): The superstructure-based (cf. chapter 4) and the superstructure-free (cf. chapter 5) synthesis methodology. Both methodologies are successfully applied to an illustrative synthesis problem. However, the illustrative problem is simplified such that it neglects any practical constraints that substantially complicate real-world problems. In this chapter, the two methodologies are explored with regard to their applicability to a real-world synthesis problem, for which already existing equipment and constructional limitations have to be taken into account.

In section 6.1, the real-world synthesis problem is formulated. Second, the synthesis problem is addressed using single-objective optimization (section 6.2). Next, the generation of near-optimal solution alternatives is discussed (section 6.3). Afterwards, multi-objective decision support is provided through the generation of Pareto-optimal compromise solutions (section 6.4). In section 6.5, the superstructure-based and the superstructure-free synthesis methodologies are evaluated with regard to the quality of the generated solutions and the required computational effort to generate these solutions. Finally, the chapter is summarized and conclusions are drawn (section 6.6).

6.1 Problem formulation

6.1.1 Site description

The case study is based on a real-world problem from the pharmaceutical industry. The considered site comprises six building complexes housing offices, and production and research facilities (Fig. 6.1). A public road separates the site into main site (A) and secondary site (B). On site A, all building complexes are connected via a central heating and cooling network. In the base case scenario, site B is not connected to the cooling, but only to the heating network. Furthermore, in base case, the production process on site B has no demand for cooling, but on site B, a new production process is installed inducing cooling demands. However, because of the public road, the installation of an additional pipe connecting site B to the cooling network on site A is not allowed. Both sites are connected to the regional natural gas grid (gas tariff: 6 ct/kWh) and the regional electricity grid (electricity tariff: 16 ct/kWh; feed-in tariff: 10 ct/kWh). Electricity generated on-site by CHP engines can either be used on-site to meet electricity demands or to run compression chillers, or else it can be fed-in to the regional electricity grid. All heat generators have to be installed on site A.

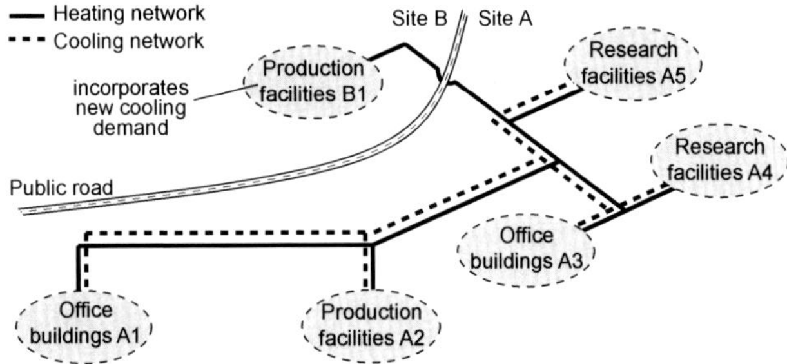

Figure 6.1: Schematic plant layout of the considered site. On site A (main site), a central heating and cooling network connects five building complexes. The building complex on site B (secondary site) is only connected to the central heating network. Establishing new connections between both sites is not possible due to the separating public road.

The described site has time-varying demands for heating, cooling, and electricity. Note that the number of periods considered during optimization strongly influences the complexity of the optimal synthesis problem due to the addition of variables in each period. Thus, the computational effort for problem solving increases nonlinearly with increasing period numbers, in the worst case even exponentially (Iyer and Grossmann, 1998; Rong et al., 2008). Therefore, in this study, monthly-averaged energy demand time series are assumed (Fig. 6.2).

The modeled demand profiles include the demands induced by the new production process on site B. The annual demands for electricity, heating, and cooling amount to 47.7 GWh, 28.1 GWh, and 27.3 GWh, respectively. The time series show a symmetric behavior around the summer months July and August. For that reason, they are

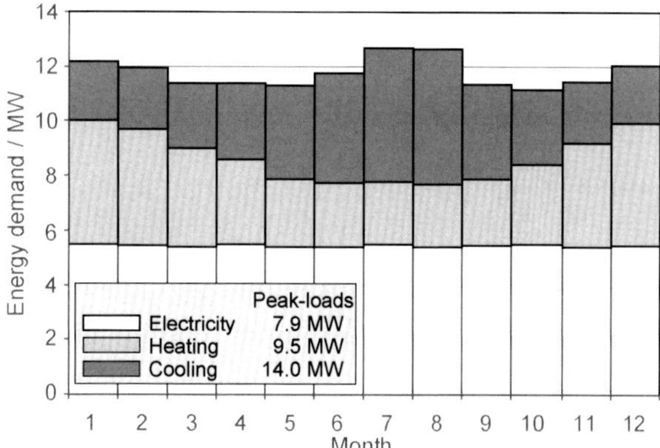

Figure 6.2: Monthly-averaged demand profiles for electricity, heating and cooling (stacked bar chart). Peak-loads (given in legend) occur only during few hours per year, and thus hardly contribute to the annual energy demands.

further simplified by aggregation of the twelve monthly-averaged time steps to only six time steps: 1) January and December, 2) February and November, 3) March, April, and October, 4) May and September, 5) June, and 6) July and August. In addition, the minimum and maximum demands are taken into account. These demands occur only during few hours per year, and thus only barely contribute to the annual energy demands. However, it is important to incorporate them in the demand profiles to guarantee adequate equipment sizing. In summary, the energy demands are modeled by eight time steps including the peak-load time steps.

The existing supply system consists of three boilers, one CHP engine, and three compression chillers (Table 6.1). However, one boiler and one compression chiller cannot be further operated, and thus require substitution.

Table 6.1: Nominal thermal powers, overall efficiencies, and COPs of the existing (E) equipment (B: boiler, CHP: CHP engine, TC: turbo-chiller).

	B E1	B E2	B E3*	CHP E1	TC E1	TC E2	TC E3*
Thermal power / MW	7.0	7.0	1.5	3.0	8.0	8.0	1.0
η_N, COP_N / -	0.8	0.75	0.75	0.8	2.8	2.8	2.4

* Boiler E3 and turbo-chiller E3 require substitution.

6.1.2 Three-step conceptual retrofit synthesis

Conceptual retrofit synthesis is performed for the described site taking into account already existing equipment and constructional limitations. For this purpose, both the superstructure-based and the superstructure-free synthesis methodologies are employed using the component models and the MILP formulation introduced in chapter 3. Commonly, optimization-based synthesis methods aim at generating the optimal solution based on a single criterion only. However, single-objective optimization models suffer from the following shortcomings: First, a mathematical model is never a perfect representation of the real world. Second, the decision maker is often not aware *a priori* of all constraints relevant for the synthesis problem at hand. Thus, the optimal solution is usually only an approximation of the optimal real-world solution. Moreover, in rapidly changing economic and technological environments (e.g., varying energy tariffs, changing energy demands, etc.), the constraints will change in the future. However, the optimal solution generally only reflects the status quo of the current situation. And finally, for real-world problems, multiple criteria (economic, environmental, social, etc.) need to be considered. However, it is generally not possible to identify a single solution that simultaneously optimizes all objectives. Due to these shortcomings, a single optimal solution does usually not suffice in practice to reach rational and far-sighted synthesis decisions. Instead, deeper understanding of the synthesis problem is required. For this purpose, a *three-step synthesis procedure* is presented that aims at providing the required understanding of the synthesis problem through the generation and analysis of a set of promising solutions rather than a single solution only:

Step 1) Generation of an economically optimal solution. As starting point, the optimal solution is identified that maximizes the net present value (NPV). Here, a cash flow time of 10 years and a discount rate of 8 % are assumed.

Step 2) Generation of near-optimal solution alternatives. A rich near-optimal solution space is expected for the considered synthesis problem (cf. section 4.3). Thus, in the second step, a ranked set of structurally different, near-optimal solutions is generated.

Step 3) Generation of Pareto-optimal compromise solutions. In the third step, multi-objective optimization is performed to provide the decision maker with information on the trade-offs between two contradiction criteria: minimization of total investments, and minimization of cumulative energy demand (CED).

Finally, the generated solution set is analyzed to identify both common features and differences among the generated candidate solutions.

6.1.3 Computer system and optimization software

All computations are performed on an Intel Xeon X5650 2.67 GHz with 2 GB RAM and four kernels. The operating system is Windows XP (32-bit). Multi-threading is activated for CPLEX to perform parallel problem solving on all four kernels. The MILP problems are implemented in the modeling language GAMS version 23.7.3. IBM ILOG CPLEX version 12.2 is used to solve the MILP problems.

For the superstructure-free synthesis methodology, the synthesis problem is decomposed into two problems for hybrid optimization (cf. section 5.2): On the upper synthesis level, an evolutionary algorithm optimizes the system structure; on the lower sizing and operation level, the system structure is fixed and deterministic MILP optimization is performed. In contrast, the superstructure-based synthesis methodology performs deterministic MILP optimization on all three synthesis levels simultaneously, i.e., system structure, sizing, and operation. For better comparison of the computational performance of both methods, the branching order of CPLEX is predefined for the solution of the MILP problems according to the hierarchical relationship of the synthesis decisions: First, the binary decision variables are branched that represent the (non-)existence of the units incorporated in the superstructure (this is the step omitted for superstructure-free synthesis); next, the binary decision variables are branched that represent the sizing of the selected units according to the piecewise linearized cost functions; finally, the binary decision variables are branched that represent the units' operation statuses.[1]

Also note that neither the solution algorithms nor the MILP formulation have been tuned to the considered synthesis problem. However, algorithm tuning and consistent application of good modeling practice, decomposition strategies, etc. have the potential to reduce computational effort by orders of magnitude as discussed in detail in the future perspective (section 7.1).

6.2 Single-objective optimal synthesis

In this section, the solution of the single-objective synthesis problem is discussed. First, the optimal solution is presented. Next, the single-objective synthesis is presented employing both the superstructure-based and the superstructure-free synthesis methodologies.

[1] Note that predefining the CPLEX-branching order to exploit the hierarchical structure of the optimization problem also enhances the solution process. For the optimization computations performed in this thesis, the solution time reductions are observed in the order of 20 %.

6.2.1 Optimal solution

Both the superstructure-based and the superstructure-free approach identify the optimal solution presented in this section (Figs. 6.3 and 6.4). This solution incorporates existing as well as new equipment. The optimal net present value adds up to −46.99 M€ (Table 6.2). This is an improvement of 39 % compared to the base case configuration, in which the additional cooling demand on site B is not even incorporated yet. The reduced connectivity matrix of the optimal solution is shown in Fig. 6.5. For clarity, technologies not selected in the optimal solution are removed from this matrix.

Table 6.2: Economic parameters of base case and NPV-optimal solution.

solution	NPV / M€	investments / M€	energy cost / M€ p.a.	maintenance cost / M€ p.a.
base case solution	−76.36	0	11.27	0.11
NPV-optimal solution	−46.99	2.35	6.44	0.22

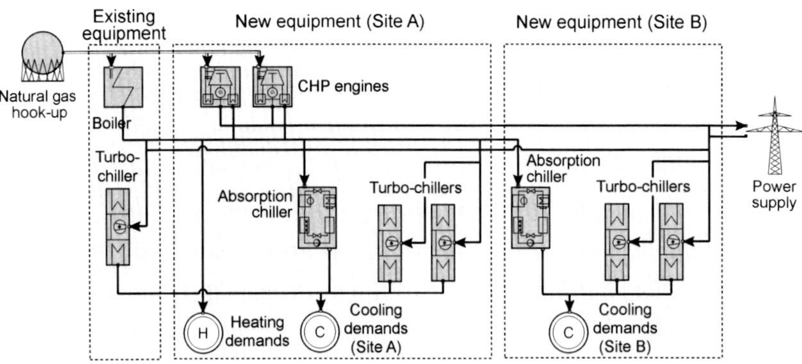

Figure 6.3: Optimal flowsheet of the real-world synthesis problem. For simplicity, the electricity demand is not shown in the figure.

In optimal configuration, the heating system consists of one already existing boiler E1 (7.0 MW) and two new equally-sized CHP engines (each 2.4 MW). CHP engine N1 is operated as central heat generator for sites A and B. In contrast, boiler E1 is reserved to cover the heating demands of building complexes A4 and B1. CHP engine N2 covers all heating demands except for those of building complexes A2 and A5.

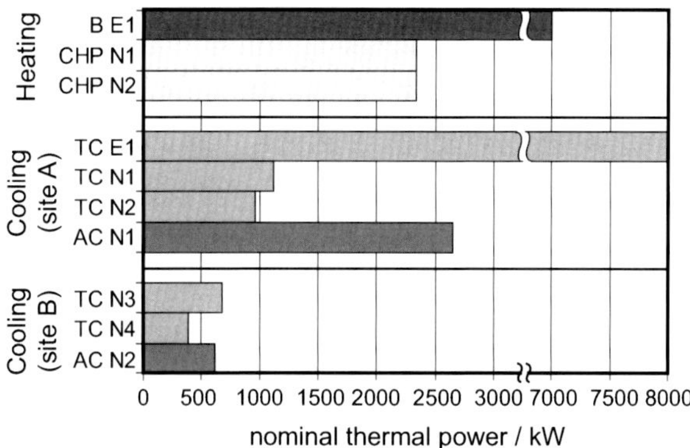

Figure 6.4: Equipment selection and sizing of the optimal solution. The bars representing the technologies' sizing are filled with different shades of gray for each type of technology.

All three heat generators provide heating for driving the absorption chillers installed on-site. Cooling on site A is supplied by the existing turbo-chiller TC1 (8.0 MW), two new compression chillers (turbo-chiller N1 and N2, each 1.0 MW), and one new absorption chiller N1 (2.6 MW). For the cooling system, absorption chiller N1, turbo-chiller N1, and turbo-chiller N2 are operated as central cooling generators for site A. Turbo-chiller E1 provides cooling for building complexes A1, A3, and A4. The cooling system on site B also encompasses two new compression chillers (turbo-chiller N3 and N4, 0.7 and 0.4 MW) and one new absorption chiller N2 (0.6 MW).

In Fig. 6.6, the annual average operating points of the installed equipment are shown: While both CHP engines are sized to enable year-round operation at full-load, the existing boiler is reserved to meet the heating peak-loads in winter at less than 50 % part-load. On site A, the existing turbo-chiller E1 is reserved to meet the cooling peak-loads in summer. The other chillers on site A (absorption chiller N1, turbo-chillers N1, and N2) are sized to allow for load sharing enabling to run them close to their maximum COPs year-round. On site B, all three chillers are operated at loads close to their maximum COPs year-round. All electricity generated by the CHP engines is used on-site for operating the compression chillers and meeting the electricity demand.

In appendix C.4, the equipment installed in the optimal solution is listed including nominal thermal powers, investment cost, operating times, and annual average part-loads.

	Demand for heating and driving heat for absorption chillers (AC)								Demand for cooling					
	A1	A2	A3	A4	A5	B1	AC1	AC2	A1	A2	A3	A4	A5	B1
Boiler E1				1		1	1	1						
CHP engine N1	1	1	1	1	1	1	1	1			0			
CHP engine N2	1		1	1		1	1	1						
Absorption chiller N1									1	1	1	1	1	
Turbo-chiller E1									1		1	1		
Turbo-chiller N1									1	1	1	1	1	
Turbo-chiller N2				0					1	1	1	1	1	
Absorption chiller N2														1
Turbo-chiller N3														1
Turbo-chiller N4														1

Figure 6.5: Reduced connectivity matrix (transposed) representing the optimal solution (E: Existing equipment. N: New equipment). For simplicity, electricity and natural gas are not shown.

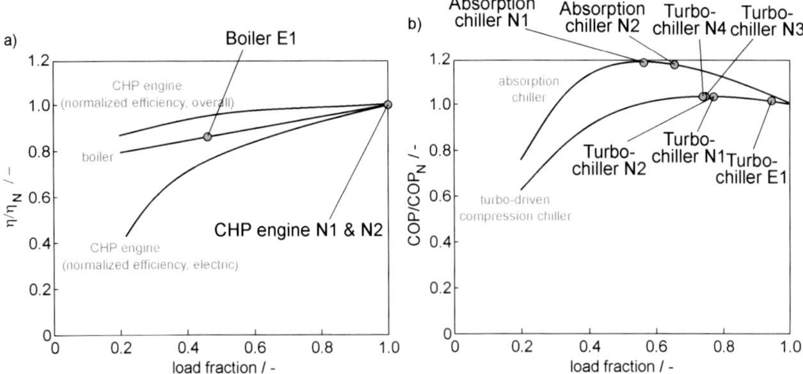

Figure 6.6: Annual average operating efficiencies (η/η_N, COP/COP_N) against operating part-loads of equipment installed in the optimal solution.

6.2.2 Superstructure-based optimal synthesis

The real-world synthesis problem is solved to global optimality (maximal relative optimality gap: 0 %). The final superstructure of the successive approach is shown in Fig. 6.7. Boiler E3 and turbo-chiller E3 require substitution, and thus are not incorporated in the superstructure.

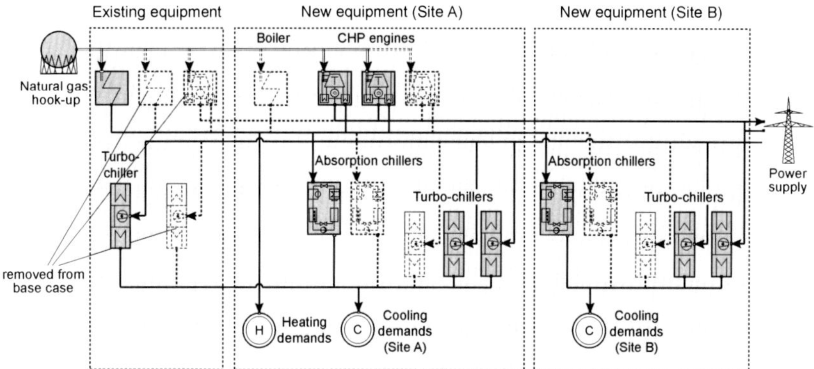

Figure 6.7: Final superstructure and optimal solution (gray units) of the real-world synthesis problem. For simplicity, the electricity demand is not shown in the figure.

To illustrate the progress of the successive optimization approach, the net present value, investment sum, and annual running cost of each solution are plotted against the optimization run number of the successive approach (Fig. 6.8). The corresponding superstructures and solutions are schematically illustrated in Fig. 6.9.

In base case scenario, the net present value amounts to −76.37 M€ (run 0: no cooling demand on site B, boiler E3 and turbo-chiller E3 still available, no investments). In run 1, the new cooling demand on site B is considered, and new equipment is incorporated in the superstructure. The corresponding solution incorporates redundant units and improves the net present value by 34.5 % (−49.96 M€). In run 2, the net present value amounts to −47.02 M€, which represents further improvement by 5.9 %. Run 3 yields the optimal solution. However, the net present value is only marginally improved to −46.99 M€. Moreover, this solution cannot be identified as optimal solution until run 4 has been performed. Only then, optimization does not further improve the current best solution, and one spare unit is available of each technology for each topographically defined site in the superstructure. Thus, the successive approach is terminated after run 4.

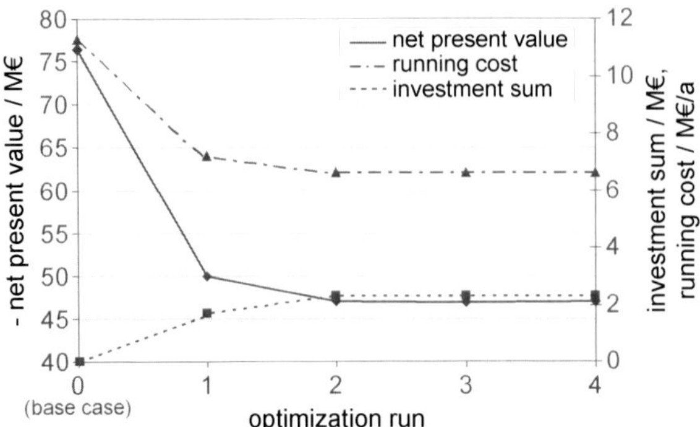

Figure 6.8: Progress of the successive optimization for the real-world synthesis problem. Net present value, investment sum, and annual running cost (energy + maintenance cost) of each optimal solution plotted against the optimization run number.

In the final solution, two turbo-chillers are installed on site B. Interestingly, however, no turbo-chiller is installed on site B in the solution of run 1. The first turbo-chiller is not installed until a second new CHP engine is installed (run 2). This is because only then does operation enable an advantageous dispatch of the CHP electricity for driving the turbo-chiller. This behavior underlines the complexity of optimal DESS synthesis problems.

In summary, the results show that the successive approach automatically identifies the optimal solution assessing the complex trade-offs between cost and number of units. Moreover, it performs automated retrofit synthesis taking into account already existing equipment and constructional limitations.

Computational performance

The combinatorial complexity of the optimal synthesis problem is tremendous. Compared to the illustrative synthesis problem (cf. section 4.3), the real-world problem is further complicated by the constructional limitation separating the site into site A and B, and by the already existing base case equipment. The model of the final superstructure (Fig. 6.7) encodes 184 320 structurally different solutions (for details on this calculation, see section 3.2.2):

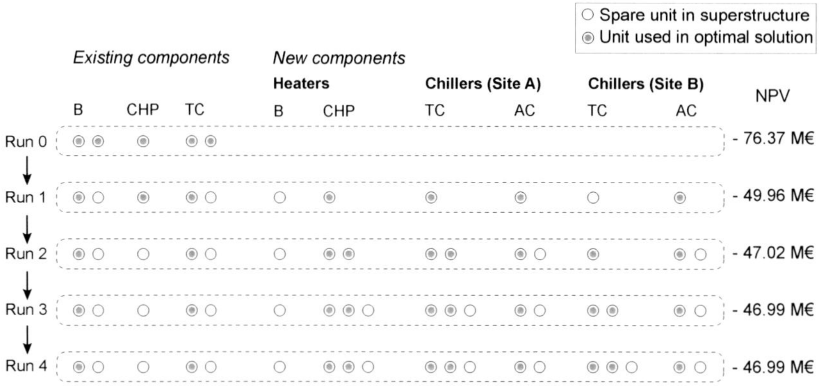

	Existing components			New components: Heaters		Chillers (Site A)		Chillers (Site B)		NPV
	B	CHP	TC	B	CHP	TC	AC	TC	AC	
Run 0	◉ ◉	◉	◉ ◉							- 76.37 M€
Run 1	◉ ○	◉	◉ ○	○	◉	◉	◉	○	◉	- 49.96 M€
Run 2	◉ ○	○	◉ ○	○	◉ ◉	◉ ◉	◉ ○	◉	◉ ○	- 47.02 M€
Run 3	◉ ○	○	◉ ○	○	◉ ◉ ○	◉ ◉ ○	◉ ○	◉ ◉	◉ ○	- 46.99 M€
Run 4	◉ ○	○	◉ ○	○	◉ ◉ ○	◉ ◉ ○	◉ ○	◉ ◉ ○	◉ ○	- 46.99 M€

○ Spare unit in superstructure
◉ Unit used in optimal solution

Figure 6.9: Progress of the successive optimization for the real-world synthesis problem. Schematic illustration of the number of units available in each superstructure and used in the optimal configuration of each optimization run (B: boiler, CHP: CHP engine, TC: turbo-chiller, AC: absorption chiller).

$$\underbrace{2^5}_{\substack{\text{base case}\\ \text{equipment}}} \cdot \underbrace{\underbrace{\binom{2+1-1}{1}}_{\text{boiler}} \cdot \underbrace{\binom{4+3-1}{3}}_{\text{CHP engines}} \cdot \underbrace{\binom{2+3-1}{3}}_{\text{turbo-chillers}} \cdot \underbrace{\binom{2+2-1}{2}}_{\text{absorption chillers}}}_{\text{site A}} \cdot \underbrace{\underbrace{\binom{2+3-1}{3}}_{\text{turbo-chillers}} \cdot \underbrace{\binom{2+2-1}{2}}_{\text{absorption chillers}}}_{\text{site B}}$$

$$= 2^5 \cdot \binom{2}{1} \cdot \binom{6}{3} \cdot \binom{4}{3} \cdot \binom{3}{2} \cdot \binom{4}{3} \cdot \binom{3}{2} = 184\,320.$$

Keep in mind that this number only reflects the combinatorial complexity for equipment selection. The complete optimization model contains further binary and continuous variables representing the decisions for equipment sizing and operation: The model of the final superstructure comprises 28 530 constraints, and 29 450 continuous and 1280 binary decision variables.

The computational effort to solve this problem to global optimality (maximal relative optimality gap: 0 %) is significant: The total number of iterations and computation time of the successive optimization amount to 231 000 000 iterations and 16:15:46 hours, respectively (Table 6.3). By far the most computations are used for the branch-and-bound algorithm to scan the search tree in order to close the optimality gap below 0.25 %. In fact, the optimal solution has already been identified at this stage, but the branch-and-bound algorithm still resolves the estimates on the other branches to prove global optimality.

Repeating the successive optimization with a maximal relative optimality gap of 0.25 % still yields the global optimal solution. At the same time, the computational effort is drastically reduced: the number of iterations and solution time amount to 29 400 000 iterations and 13:47 minutes, respectively (Table 6.3). In fact, optimization with this setting yields identical solutions for all steps of the successive optimization. If larger optimality gaps are accepted, optimization identifies only suboptimal solutions.

Table 6.3: Computational effort of the successive optimization to identify the optimal solution of the real-world synthesis problem with relative maximal optimality gaps of 0 % and 0.25 %.

run	NPV / M€	relative optimality gap	optimality gap: 0 %		optimality gap 0.25 %	
			number of iterations / 10^6	solution time / hh:mm:ss	number of iterations / 10^6	solution time / hh:mm:ss
0	−76.369	62.5 %	< 1	00:00:01	< 1	00:00:01
1	−49.956	5.90 %	< 1	00:00:12	< 1	00:00:07
2	−47.018	0.10 %	7	00:06:42	3	00:01:45
3	−46.986	0	37	01:53:33	11	00:04:58
4	−46.986	0	96	14:35:11	16	00:06:56
		$\sum$	141	16:15:46	29	00:13:47

6.2.3 Superstructure-free optimal synthesis

For superstructure-free synthesis, the hybrid optimization algorithm introduced in section 5.2 is employed. For optimization on the synthesis level, the hybrid algorithm uses an evolutionary algorithm that performs structural mutation based on the extended energy conversion hierarchy encompassing boilers, CHP engines, compression, and absorption chillers (cf. section 5.4.2). The lower level sizing and operation problems are solved by MILP optimization to global optimality (maximal relative optimality gap: 0 %). The $(\mu + \lambda)$-evolutionary algorithm is parameterized based on the results of a simple parameter study: The selection parameters are set to $\mu = 5$ and $\lambda = 20$; i.e., in each generation, 20 offspring individuals are generated from 5 (current best) parent individuals. The 5 best individuals of each generation remain in the population for the next generation. The employed production set and the production probabilities are listed in appendix B.3. The mutation strength, i.e., the number of productions performed within one mutation step, is randomly chosen for each generation according to a uniform probability distribution. The minimum mutation strength is set to one. The maximum mutation strength is set to five. The evolutionary optimization is terminated after 50 generations, i.e., 1000 individuals are evaluated.

The provided initial solution incorporates the already installed equipment from the the base case scenario (Table 6.1) and one new turbo-chiller providing cooling energy to site B. Boiler E3 and turbo-chiller E3 are not available in the initial solution, which in turn is not feasible because the existing boilers and turbo-chillers are too large to be operated at the part-loads necessary to meet the minimum loads.

The superstructure-free synthesis is performed ten times (Table 6.4). Eight of the ten runs identify the global optimal solution. The other two runs generate only suboptimal, however near-optimal, solutions with objective function values that differ only 0.08 % and 0.07 % from the optimal objective function value.

Table 6.4: Ten runs of the superstructure-free synthesis method. Listed are the generation numbers, in which the optimal solution has been identified, the solution times required to identify the optimal solution, the overall solution times, and the objective function values (NPV) with corresponding relative optimality gaps of the best solutions found. Times given in hh:mm:ss.

run	generation number	solution time / hh:mm:ss to optimality	solution time / hh:mm:ss overall	NPV / M€	relative optimality gap
1	12	00:31:52	03:07:41	−46.986	0
2	16	01:34:02	06:42:31	−46.986	0
3	17	00:29:47	02:32:03	−46.986	0
4	17	00:46:06	04:11:02	−46.986	0
5	18	00:42:23	03:58:43	−46.986	0
6	23	03:43:33	06:44:52	−46.986	0
7	31	01:18:48	04:18:26	−46.986	0
8	37	05:47:13	07:13:12	−46.986	0
9	50	-	04:09:17	−47.025	0.08 %
10	50	-	06:20:11	−47.018	0.07 %

In terms of performed generations, the fastest run identifies the optimal solution already within the first 12 generations. The computation times vary between 2:32:03 hours and 7:13:12 hours with an average solution time of about 5 hours. The computation times to generate the optimal solution vary between 29:47 minutes and 5:47:13 hours. No correlation is observed between the computation time and the quality of the best solution found. However, if many complex intermediate solutions are generated, optimization usually takes longer than for more direct optimization routes with less complex structures.

For the superstructure-based synthesis approach, the maximal relative optimality gap has been set to 0.25 % to reduce the solution time to acceptable values. Thus, for a second series of ten optimization runs with the superstructure-free approach, the maximal relative optimality gap is also set to 0.25 % for the solution of the MILP subproblem. With this setting, the solution times are reduced to on average 3 hours (between 1:50:04 hours and 4:02:21 hours) without deteriorating the solution quality.

6.3 Generation of near-optimal solution alternatives

The successive optimization generates two practically equally good solutions, i.e., the optimal solution and one near-optimal solution (run 2). This indicates a rich near-optimal solution space, which is thoroughly explored in the following by both synthesis methodologies.

6.3.1 Superstructure-based alternatives generation

To generate structurally different, near-optimal solution alternatives, the superstructure-based method uses the integer-cut approach (cf. section 4.3.4). To limit computational effort, the successive superstructure expansion strategy is not applied within the integer-cut approach. Instead, the integer-cut approach is applied to the final superstructure model of the successive approach, thus limiting the identified solutions to those embedded in the final superstructure. The maximal relative optimality gap is set to 0.25 % (cf. section 6.2.2).

A manageable set of ten solution structures is generated (Table 6.5). The objective function values of the ten best solutions lie within an optimality gap of 0.17 %. Thus, the generated solutions are practically equally good.

The total number of iterations and solution time amount to 207 000 000 iterations and 1:55:35 hours, respectively (Table 6.6). The required number of iterations and solution time grow roughly proportionally with the number of generated solutions.

Table 6.5: The ten best solution structures of the real-world problem. Listed are the objective function values (NPV), the relative optimality gaps, and the solution structures. The solution structures are given in numbers of installed equipment (B: boiler, CHP: CHP engine, TC: turbo-chiller, AC: absorption chiller).

k-th best solution	NPV / M€	relative optimality gap	solution structure: heaters B	heaters CHP	site A TC	site A AC	site B TC	site B AC
1st	−46.986	0	1	2	3	1	2	1
2nd	−47.018	0.07 %	1	2	3	1	1	1
3rd	−47.018	0.07 %	1	2	2	1	2	1
4th	−47.024	0.08 %	2	2	3	1	2	1
5th	−47.033	0.10 %	1	2	3	2	1	1
6th	−47.035	0.10 %	1	2	3	2	2	1
7th	−47.041	0.12 %	1	2	3	1	3	1
8th	−47.047	0.13 %	1	2	3	1	1	2
9th	−47.056	0.15 %	2	2	3	1	1	1
10th	−47.068	0.17 %	1	2	3	2	2	0

Table 6.6: Computational effort to generate the ten best solutions of the real-world synthesis problem. Maximal relative optimality gap: 0.25 %. Solution time given in mm:ss.

k-th best solution	1st*	2nd*	3rd	4th	5th	6th	7th	8th	9th	10th
iterations / 10^6	29.4	–	28.5	23.5	17.5	22.3	19.2	26.4	21.2	17.7
solution time	13:47	–	12:41	14:39	09:58	17:44	08:35	14:09	12:58	11:04

* The computational effort for generating the two best solutions includes all computations of the successive optimization.

6.3.2 Superstructure-free alternatives generation

An important feature of the superstructure-free approach is that it generates intermediate solutions during the search for the optimal solution. In the following, optimization run 5 (cf. section 6.2.3) is examined in more detail as exemplary run (Fig. 6.10). For the other optimization runs, a similar convergence behavior is observed as for the described exemplary run 5.

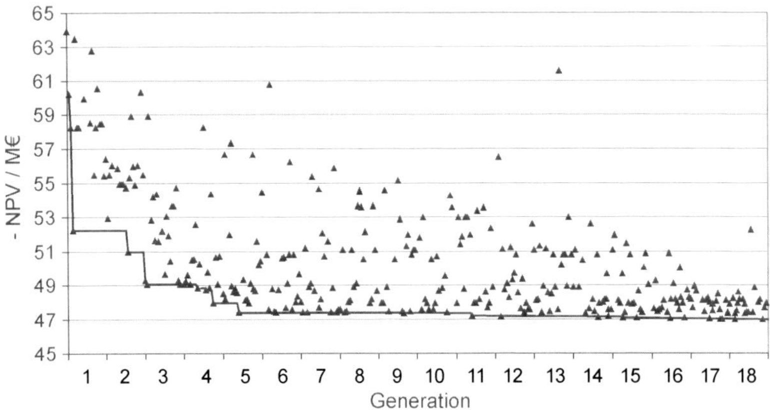

Figure 6.10: Progress of optimization run 5 until the optimal solution is identified (generation 18). Net present value (NPV) plotted against generation number. The current best solutions are illustrated by the straight lines.

The superstructure-free synthesis methodology identifies good solutions already within the first few generations. However, with proceeding generations, the improvements become smaller, and the current best solution is only gradually improved until the optimal solution is identified. Simultaneously, many poor solutions are generated during the whole optimization process. However, as for the improvements of the objective function values, also the number of poor outliers shrinks with increasing generation numbers: This is due to the fact that more and more parent individuals lie within the near-optimal solution space, and thus very large, disadvantageous mutation steps become necessary to generate such poor outliers. Note that poor outliers will never completely vanish from the optimization, but will always be generated now and then. This behavior reflects the important feature of the proposed approach guaranteeing that the search will not get stuck in suboptimal local solutions.

Table 6.7: Near-optimal intermediate solutions of optimization run 5 (until generation 18). The solutions are sorted by individual and generation numbers. The solution structures are given in numbers of installed equipment (B: boiler, CHP: CHP engine, TC: turbo-chiller, AC: absorption chiller).

individual	generation	NPV / M€	relative optimality gap	solution structure					
				heaters		site A		site B	
				B	CHP	TC	AC	TC	AC
175	9	−47.457	1.00 %	2	2	2	2	0	1
195	10	−47.401	0.88 %	1	2	3	1	0	1
206	11	−47.226	0.51 %	1	2	2	1	0	2
223	12	−47.122	0.29 %	2	2	2	1	1	2
275	14	−47.078	0.20 %	1	2	3	2	3	0
279	14	−47.075	0.19 %	1	2	4	2	3	0
291	15	−47.069	0.18 %	1	2	3	2	1	2
308	16	−47.069	0.18 %	1	2	2	2	1	1
316	16	−47.065	0.17 %	1	2	4	1	3	1
331	17	−47.044	0.12 %	1	2	4	1	1	1
338	17	−47.035	0.10 %	1	2	3	2	2	1
339	17	−47.025	0.08 %	1	2	4	1	2	1
342	18	−47.018	0.07 %	1	2	2	1	2	1
358	18	−46.986	0	1	2	3	1	2	1

During optimization run 5, 34 structurally different, near-optimal solutions are generated with objective function values that lie within a relative optimality gap of 1 %. In the following, only those 14 near-optimal solutions are discussed that are generated within the 18 generations required to identify the optimal solution (Table 6.7). Three of these 14 solutions are among the ten best-ranked solutions identified by the integer-cut approach (individuals 338, 342, and 359, cf. section 6.5). Among the remaining eleven near-optimal solutions, there are some solution structures that differ very much from those identified by the integer-cut approach. For example, individuals 279, 316, 331, and 339 incorporate four turbo-chillers on site A. Moreover, there is a larger diversity of candidate solutions that incorporate either only absorption chillers or only turbo-chillers on site B (individuals 175, 195, 206, 275, and 279).

6.3.3 Discussion of the near-optimal solution alternatives

The NPVs of the generated near-optimal solution alternatives lie within a relative optimality gap of 1 %. Considering the multitude of additional constraints and uncertainties arising in practice (e.g., cost for equipment installation and control, flexibility towards changing demands, varying energy prices, etc.), it is practically impossible to make a clear statement about which solutions are better than others strictly based on the NPV. Thus, for rational synthesis decisions, deeper insight is required into the features of the generated solutions: For this purpose, first, running cost are plotted against total investments of the generated near-optimal solution alternatives (Fig. 6.11). The plot supports the decision maker to choose between solution alternatives requiring lower investments at the expense of higher running cost, or the other way round. For the considered synthesis problem, the total investments of the near-optimal solution alternatives vary at maximum by 400 000 € (16 %). The running cost vary at maximum by 110 000 €/a (1.7 %).

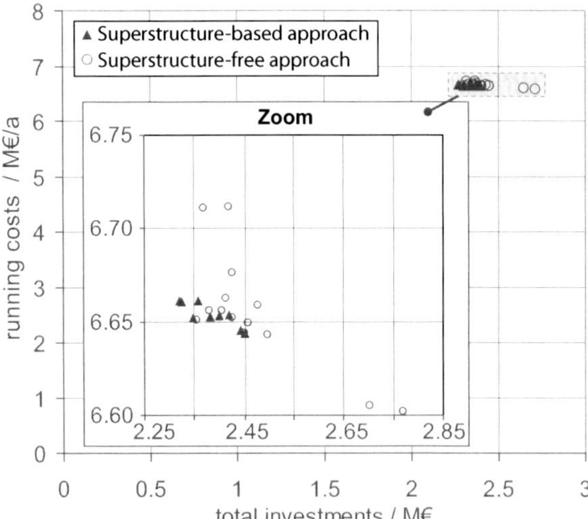

Figure 6.11: Running cost plotted against total investments of all near-optimal solution alternatives generated by the superstructure-based (cf. section 6.3.1) and the superstructure-free (cf. section 6.3.2) synthesis methodologies.

Secondly, the near-optimal solution alternatives are compared with regard to equipment configuration and sizing: As common feature, the already existing boiler E1 and turbo-chiller E1 remain in all near-optimal solution alternatives (for meeting peak-load demands). Furthermore, all near-optimal solution alternatives incorporate exactly two CHP engines. Besides these common features, the generated solution alternatives differ with respect to the remainder equipment in both configuration and sizing. In the following, prominent solution alternatives are discussed that feature special structural characteristics (Fig. 6.12):

- Solutions that incorporate only few units are interesting options because they minimize the complexity of equipment installation and control: E.g., the 3rd best solution contains only two compression and one absorption chiller on site A; the 5th, 9th, and 10th best solutions install only two chillers on site B (cf. Table 6.5 and Fig. 6.12 a), c), d)). Moreover, individuals 175 and 195 incorporate only one absorption chiller on site B (cf. Table 6.7 and Fig. 6.12 e)).
- On the other hand, solutions that incorporate many units are interesting alternatives because they provide greater flexibility with regard to decentralization options, operation strategies, system up- or down-scaling, etc.: E.g., the 4th and 9th best solution install two boilers, and the 7th best solution installs three turbo-chillers on site B (cf. Table 6.5 and Fig. 6.12 b), c)). Moreover, individuals 279, 316, 331, and 339 incorporate up to four redundant units (cf. Table 6.7 and Fig. 6.12 f)).
- Solutions that do not incorporate any absorption chillers on site B avoid costs related to the heating network installation and operation: This is the case, e.g., for the 10th best solution (cf. Table 6.5 and Fig. 6.12 d)), and individuals 275 and 279 (cf. Table 6.7 and Fig. 6.12 f).
- Finally, individuals 175, 195, and 206 do not incorporate any compression chillers on site B, thus maximizing trigeneration (cf. Table 6.7 and Fig. 6.12 e).

The generation and analysis of the near-optimal solution alternatives supports the decision maker to identify both the common features and the differences of the generated solutions. The former can be considered as "must haves" of good solutions. The latter support the decision maker to account for aspects that have not been explicitly considered during optimization. In summary, this approach provides deeper understanding of the synthesis problem and wider space for rational synthesis decisions.

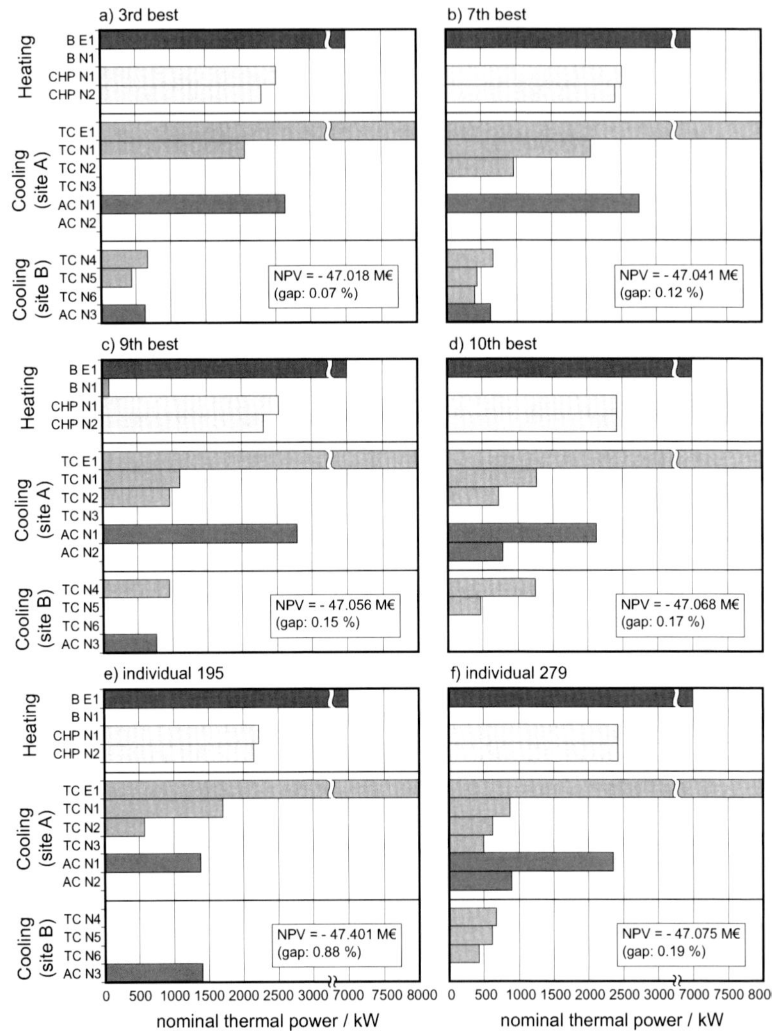

Figure 6.12: Equipment selection and sizing for six near-optimal solution alternatives. a)-d) superstructure-based approach: a) 3rd best, b) 7th best, c) 9th best, d) 10th best; e)-f) superstructure-free approach: e) individual 195, f) individual 279. The bars representing the technologies' sizing are filled with different shades of gray for each type of technology.

6.4 Multi-objective optimal synthesis

Single-objective optimization minimizing the net present value (NPV) reduces the synthesis task to the identification of the economically optimal solution. However, using the NPV as optimization criterion causes to assign fixed weights to the required investments and running expenses according to the assumed cash flow time and discount rate, here 10 years and 8 %, respectively. This approach reflects only the status quo of the current situation, but increasing energy prices, e.g., will lead to situations where more energy-efficient and more capital-intensive candidate solutions will outmatch cheaper and less energy-efficient solutions. Moreover, for economic optimization, ecological considerations are neglected, which gain in importance in engineering practice. Thus, in this section, a more complete picture of the synthesis problem is provided by multi-objective optimal synthesis. The optimization is performed with regard to two conflicting objectives: minimization of total investments and minimization of annual cumulative energy demand (Huijbregts et al., 2010). The cumulative energy demand (CED) represents the total primary energy use for the generation of a product, here heating, cooling, and electricity, taking into account the relevant front-end process chains (crude oil processing, energy transport, etc.). For energy supply systems, the CED represents a measure for the system's overall energy efficiency (Röhrlich et al., 2000). For the calculation of the CED, the following *CED-factors* are assumed: 2.96 kWh_{CED} /kWh_{el} and 1.11 kWh_{CED} /kWh_{gas} for electricity and natural gas, respectively (Fritsche and Schmidt, 2007).

Both the superstructure-based and the superstructure-free synthesis methodologies are used to generate Pareto-fronts of the multi-objective optimization problem. The superstructure-based approach implements the ε-constraint method (section 6.4.2). The superstructure-free optimization approach uses an aggregation selection technique based on the SMS-EMOA (section 6.4.3). In accordance with the single-objective optimal synthesis, the MILP problems are solved to optimality with a maximal relative optimality gap of 0.25 %.

6.4.1 Pareto-optimal solutions

The generated Pareto-front is shown together with the NPV-optimal solution in Fig. 6.13. For better comparison of the Pareto-optimal solutions and the NPV-optimal solution, in this plot, only equipment structure and sizing are retained from the NPV-optimal solution, but operation is optimized with regard to CED. The Pareto-front spans a solution space from 0.41 M€ to 6.24 M€ with respect to total investments and from 220.6 GWh to 82.4 GWh with respect to annual CED. In other words, the Pareto-optimal solutions vary by factors 15 and 2.8 with regard to total investments and annual CED, respectively. Keep in mind that solutions without any investments are not feasible due to the new production process installed on site B that requires cooling, for which new chillers have to be installed.

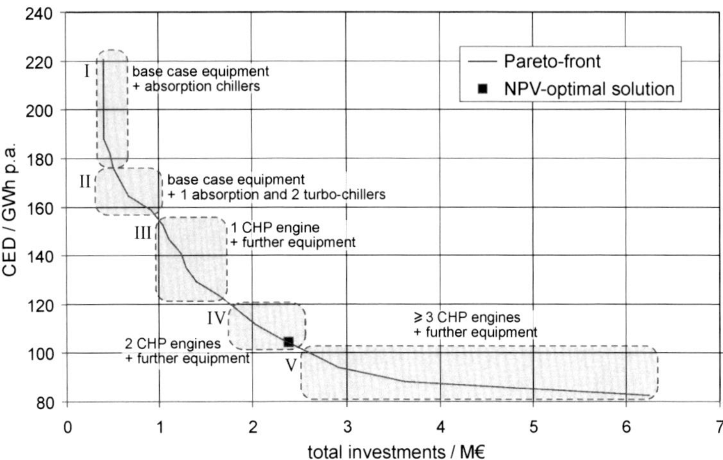

Figure 6.13: Pareto-front classified into five groups plotted together with the NPV-optimal solution.

Analysis of the Pareto-solutions shows that the Pareto-front can be structured into five groups. Within each group, the solutions are similar with respect to equipment selection, however, they differ in equipment sizing:

- In the solutions of group I, all base case equipment is kept, thus enabling to meet the energy demands at minimum investments; furthermore, up to two absorption chillers are installed: one on site A and one on site B.
- In the solutions of group II, except for one boiler, all base case equipment is kept. Moreover, two new turbo-chillers (one on each site) and one new absorption chiller (on site B) are installed. The transition from group I to group II marks the break-even point, at which it is beneficial to install turbo-chillers in addition to absorption chillers.

Groups III, IV, and V incorporate one, two, and three or more CHP engines, respectively. The number of base case equipment employed in the solutions decreases from groups III to V, i.e., with decreasing CED. The numbers of newly installed turbo-chillers and absorption chillers depend on the number and sizing of the installed CHP engines, and thus vary from group to group, but also within the groups:

- In group III, all base case equipment is kept except for one boiler. In addition, one new turbo-chiller and one absorption chiller are installed on site A, and one absorption chiller is installed on site B.
- In group IV, only one boiler and one turbo-chiller are retained from the base case to meet heating and cooling peak-loads. Moreover, on both sites, up to two new turbo-chillers and one absorption chiller are installed.
- In group V, no equipment is kept from the base case. The CED-optimal solution incorporates five new CHP engines, four absorption chillers on site A, and three absorption chillers on site B.

Extreme trade-offs appear in the regions of the Pareto-front that require particularly low total investments (group I) and particularly low CED (group V): Within group I, e.g., accepting marginal higher total investments achieves considerable savings with respect to CED; the same is valid the other way round for group V. In contrast, in group IV, which includes the NPV-optimal solution and all of the generated near-optimal solution alternatives (cf. section 6.3.3), only moderate trade-offs can be achieved. In fact, significant CED-savings require significant higher investments for the installation of a third CHP engine. On the other hand, significant investment reductions can be achieved only by sparing one of the two CHP engines, which causes the CED to increase by at least 20 %.

In summary, the generated Pareto-front supports the decision maker to evaluate the impact of additional expenditures on possible energy savings, or the other way round. Furthermore, the Pareto-front enables to analyze the impact of equipment selection and sizing on these trade-offs.

6.4.2 Superstructure-based multi-objective optimization

The superstructure-based optimization method implements the ε-constraint method (cf. section 2.4) as presented by Mavrotas (2009): The reformulated single-objective optimization problem minimizes total investments while setting upper bounds on the CED as constraints. First, single-objective optimization is performed minimizing each objective function. The results of these computations are listed in the so-called *payoff table*. The payoff table defines the ranges of the objective function values that will be covered by the ε-constraint method. For the real-world problem, the payoff table (Table 6.8) lists one solution with minimum total investments and corresponding CED, and one solution with minimum CED and corresponding total investments. Next, the CED-range is discretized. To generate the Pareto-front, one single-objective optimization problem is solved for each discretization node assuming the nodes as upper bounds on the CED. For each single-objective optimization, the successive approach is used for automated superstructure generation and expansion.

Table 6.8: Payoff table of the two objective functions total investments (f_1) and annual CED (f_2).

	f_1 / M€	f_2 / GWh p.a.
min f_1	0.41	220.55
min f_2	6.24	82.44

For the considered optimization problem, the CED-range is discretized by 25 equally-spaced nodes. The generated Pareto-front (Fig. 6.14) equals the one discussed in the previous section. The generation of the Pareto-front is computationally involved: The total number of iterations and solution time amount to 682 000 000 iterations and 14:10 hours, respectively. More than 90 % of the computational effort can be attributed to the generation of the three CED-best solutions (634 000 000 iterations and 13:12 hours), which incorporate many pieces of equipment to enable optimal energy-efficient operation. However, in practical applications, the decision maker will be mainly interested in solutions that lie in the direct neighborhood of the NPV-optimal solution. Thus, in the following, multi-objective optimal synthesis is performed once more; but this time, only the direct neighborhood of the NPV-optimal solution is considered. For this purpose, upper bounds are set on the objective functions limiting them to 20 % above the corresponding values of the NPV-optimal solution. For comparison, the CED-range is again discretized by 25 equally-spaced nodes. The generated *constrained Pareto-front* (Fig. 6.14) allows for the same analysis regarding the equipment installed in the Pareto-optimal solutions and the trade-offs between total investments and CED as discussed in section 6.4.1, but only in the direct neighborhood of the NPV-optimal solution. However, the computational effort to generate the constrained Pareto-front is drastically reduced to 49 000 000 iterations and 39:45 minutes solution time. Thus, if only the direct neighborhood of the NPV-optimal solution is considered, and in particular, if solution structures with an excessive number of units are neglected (as is the case here), the proposed method can efficiently generate a Pareto-front for the analysis of the trade-offs between total investments and CED.

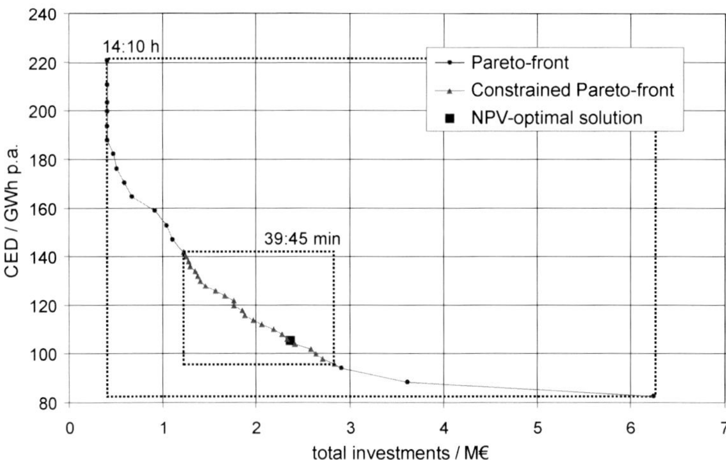

Figure 6.14: Complete Pareto-front and constrained Pareto-front in NPV-optimal neighborhood.

6.4.3 Superstructure-free multi-objective optimization

The superstructure-free approach implements an aggregation selection technique based on the SMS-EMOA for multi-objective optimization (cf. section 2.4). For aggregation selection, the weighting method is used with weights $w_k \in [0, 1]$ for each objective function k such that $\sum_k w_k = 1$. The weights are randomly varied from individual to individual. The population consists of $\mu = 25$ parent individuals and $\lambda = 1$ offspring individual. Parameter μ is chosen according to the results of a simple parameter study. For optimization, 2000 generations are computed. As initial solution, the NPV-optimal solution is provided (cf. section 6.2.2).

For most parts, the Pareto-front generated by the superstructure-free approach (Fig. 6.15) is congruent with the Pareto-front generated by the superstructure-based approach (cf. previous section 6.4.2). Moreover, five solution clusters are visible, which correspond to the five solution groups generated by the ε-constraint method: The solutions within the different clusters incorporate similar equipment. The compromise solutions within each cluster differ in equipment sizing.

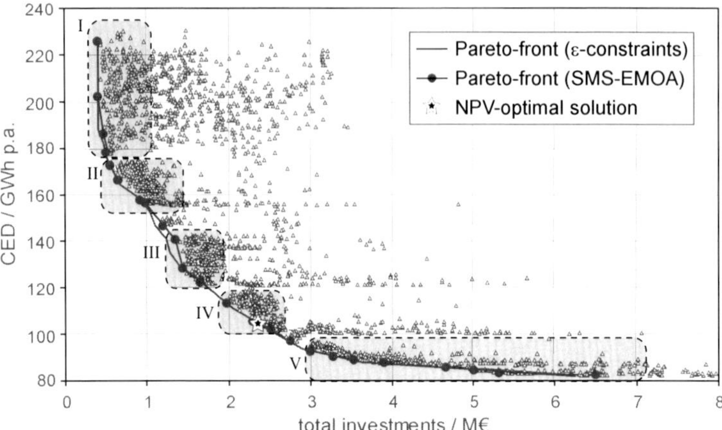

Figure 6.15: Results of the aggregation selection SMS-EMOA: Pareto-front and dominated individuals with five solution clusters plotted together with the NPV-optimal solution and the ε-constraint Pareto-front.

To illustrate the progress of the evolutionary Pareto-front generation, the individuals generated in generation 250, 500, and 2000 are plotted in Fig. 6.16. The corresponding solution times to generate these solutions are 22 minutes, 43 minutes, and 3:02 hours, respectively. Already in generation 250, the clustering of the generated individuals is recognizable, thus enabling to analyze the trade-offs between investments and CED for structurally different solutions. Among the generated individuals, there are Pareto-optimal representatives of clusters II, III, IV, and V. However, compared to the ε-constraint Pareto-front, in particular the extreme solutions with minimum total investments and minimum CED are not yet well represented. From generation 250 to 500, the approximation of the ε-constraint Pareto-front is further improved, especially in the region with minimum total investments (cluster I). Moreover, clustering of the individuals becomes more distinct, and the individuals are more widespread within the clusters. From generation 500 to 2000, the Pareto-front is further improved in the extreme objective function regions. Furthermore, in the regions between clusters II and III, and between IV and V, sub-clusters emerge that further approximate the ε-Pareto-front. In addition, a more representative distribution of individuals is generated within the clusters, thus enabling to analyze the trade-offs between investments and CED not only for structurally different solutions, but also for structurally similar solutions with different equipment sizing.

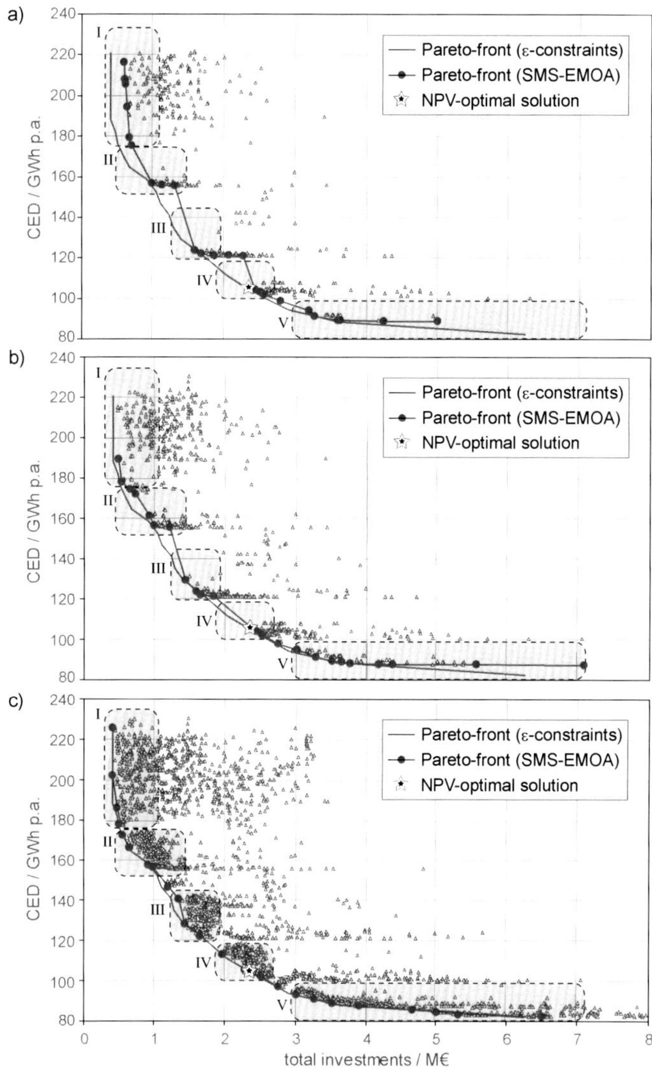

Figure 6.16: Pareto-fronts and dominated individuals plotted together with the ε-constraint Pareto-front and the NPV-optimal solution: a) generation 250 (22 min solution time), b) generation 500 (43 min solution time), c) generation 2000 (3:02 h solution time).

In summary, it is shown that the superstructure-free approach is suited to support multi-criteria decision making by generating Pareto-fronts for the analysis of the trade-offs between total investments and CED due to different equipment selection and sizing. Moreover, already in an early stage of the optimization process, the EMOA approach is capable to generate Pareto-optimal compromise solutions that differ in equipment selection. However, if the complete picture of trade-offs is desired including different equipment selection and sizing over the whole range of objective function values, more extensive optimization runs become necessary; for the considered example, up to 2000 generations.

6.5 Comparative evaluation

In the following, the synthesis methods proposed in this thesis are evaluated with regard to practicality for real-world synthesis problems. For this purpose, the solution quality and computational performance of both methods are compared for the considered synthesis problem.

6.5.1 Solution quality

The solution of the real-world synthesis problem demonstrates that both methods can solve retrofit problems of practical size while taking into account already existing equipment and constructional limitations. To identify the economically optimal solution minimizing the net present value (NPV), the methods balance capital cost against running expenses to trade-off the possibilities of keeping existing (possibly low-performance) equipment against purchasing new (high-performance) equipment. Obviously, this is a complex task, for which the methods identify and evaluate a great many options including conventional, cogeneration, and trigeneration concepts. For the considered real-world problem, both methods are capable to identify the NPV-optimal solution, however, only the superstructure-based approach guarantees optimality.

Moreover, both methods thoroughly explore the rich near-optimal solution space. The superstructure-based approach generates the ten best near-optimal solutions encoded in the final superstructure model of the successive optimization. The superstructure-free approach generates more near-optimal solutions than the superstructure-based approach. Moreover, these solutions are characterized by a larger diversity with regard to equipment selection than the solutions generated by the superstructure-based approach. However, the superstructure-free approach does not necessarily generate the best among the near-optimal solution alternatives.

Finally, both methodologies perform multi-objective optimization with regard to two contradicting objectives: minimization of total investments and minimization of annual cumulative energy demand (CED). The Pareto-fronts generated by both methods are practically congruent. They provide insight into the dependencies of the trade-offs between total investments and CED on equipment selection and sizing.

In summary, both approaches are capable of performing single- and multi-objective optimization-based synthesis, thus generating the optimal solution, a set of near-optimal solution alternatives, and a set of Pareto-optimal compromise solutions. Optimality, however, is guaranteed only by the superstructure-based approach. On the other hand, the superstructure-free approach generally generates more solution alternatives and a larger diversity of structurally different solutions.

6.5.2 Computational performance

In the following, both methodologies are compared with regard to the computational effort required for solving the synthesis tasks. For better comparison, the computational effort of the performed optimization computations is also listed in Table 6.9.

If the superstructure-based synthesis methodology is employed to solve the considered single-objective synthesis problem to global optimality (maximal relative optimality gap: 0 %), the computational effort is significant (solution time over 16 hours). On the other hand, a maximal relative optimality gap of 0.25 % still enables the superstructure-based approach to identify the global optimal solution while the solution time can be reduced to 13:47 minutes. The computational effort for generating a ranked set of structurally different near-optimal solution alternatives increases proportionally to the number of solutions to be generated. With the maximal relative optimality gap of 0.25 %, the integer-cut approach requires 1:55 hours to generate the ten best, structurally different solutions.

The superstructure-free synthesis methodology computes 50 generations for single-objective optimization. This equals the generation and evaluation of 1000 individuals. If the MILP subproblem is solved to global optimality, optimization takes on average about 5 hours. If the maximal relative optimality gap is loosened to 0.25 %, the average computation time is reduced to about 3 hours. This method cannot guarantee global optimality, but it does not require additional computations to generate near-optimal solution alternatives. In fact, in all performed optimization runs, the superstructure-free approach identifies many near-optimal solutions already within computation times below 3 hours. Often, even the optimal solution is identified within less than 2 hours. Moreover, it should be pointed out that the solution time of the superstructure-free approach does not depend on the maximal optimality gap as strongly as the super-

Table 6.9: Overview of the computational effort for the performed optimization computations (max. gap = maximal relative optimality gap).

max. gap	Superstructure-based			Superstructure-free	
	iterations / 10^6	solution time	comment	solution time	comment
Section 6.2: Single-objective optimal synthesis					
0 %	141	≈ 16 hours	-	≈ 5 hours for 1000 individuals	8 of 10 runs identify the optimum
0.25 %	29	≈ 14 min	-	≈ 3 hours for 1000 individuals	-
Section 6.3 Generation of near-optimal solution alternatives					
0.25 %	207	≈ 2 hours	10 best solutions within 0.17 % optimality gap	no extra computations required	34 solutions within 1 % optimality gap
Section 6.4: Multi-objective optimal synthesis					
0.25 %	682	≈ 14 hours	complete Pareto-front	≈ 3 hours for 2000 individuals	complete Pareto-front
0.25 %	49	≈ 40 min	constrained Pareto-front (NPV-optimal region)	≈ 22 min for 250 individuals	clustered Pareto-front (trade-offs of different structures only)

structure-based approach does. This can be attributed to the decomposition approach that performs deterministic optimization only for the much simpler sizing and operation subproblem.

Generating the complete Pareto-front using the superstructure-based approach takes about 14 hours and requires 682 000 000 iterations if the maximal relative optimality gap is set to 0.25 %. In contrast, the superstructure-free approach requires only about 3 hours for performing 2000 generations while yielding practically the same results. However, if for superstructure-based approach is employed to generate Pareto-optimal solutions in the direct neighborhood of the NPV-optimal solution only (investments and CED are limited to values of 20 % above the corresponding values of the NPV-optimal solution), the computational effort to generate this constrained Pareto-front is reduced by more than 90 % to 49 000 000 iterations and 39:45 minutes solution time. Then again, the superstructure-free approach yields a clustered approximation of the Pareto-set incorporating structurally different solutions already in generation 250 (22 minutes solution time).

To conclude the discussion of the computational performance, the computational effort for solving the single-objective problem is evaluated for both methodologies with regard to the combinatorial complexity of the considered synthesis problem: The final superstructure model of the successive approach encodes 184 320 solution structures (cf. section 6.2.2). According to the solution times of single-objective superstructure-free optimization, the sizing and operation optimization takes on average 18 or 11 seconds for a single solution structure, depending on whether the maximal relative optimality gap is set to 0 % or 0.25 %, respectively. Thus, if all solution structures were individually evaluated, the optimization would take more than 38 or 23 days, respectively. These numbers underline the computational efficiency of both the superstructure-based and the superstructure-free methods, which require only fractions of these times (about 16 hours or 14 minutes for the superstructure-based approach, and about 5 or 3 hours for the superstructure-free approach).

6.5.3 Guidelines for the selection of a synthesis methodology

Both methodologies enable automated optimal synthesis of distributed energy supply systems on the conceptual design level. However, only the superstructure-based synthesis methodology guarantees optimality. Hence, if only a few, including the best, solution alternatives have to be generated, the superstructure-based approach is to be preferred. On the other hand, the solution time of the superstructure-based synthesis methodology increases proportionally to the number of solution alternatives to be generated. Thus, if the main interest is on the generation of many near-optimal solution alternatives, the superstructure-free synthesis methodology is equally suitable or even more suited than the superstructure methodology; i.e., the solution time for the generation of the same number of near-optimal solution alternatives is in the same order of magnitude or even smaller for the superstructure-free approach, depending on the concrete number of solutions to be generated.

If, for multi-objective optimization, a complete picture is desired of the trade-offs between total investments and annual CED over the whole range of objective function values, the superstructure-free approach is most suited. On the other hand, if only the direct neighborhood of the NPV-optimal solution is of interest, then the superstructure-based method is favored to generate a constrained Pareto-front around the NPV-optimal solution.

Keep in mind that these guidelines can be considered only as rules of thumb due to the fact that the exact numbers of the solution times strongly differ depending on the employed mathematical programming formulation and the assumed maximal optimality gap. Moreover, it should be noted that consistent application of good modeling practice, decomposition strategies, etc. have the potential to reduce computational effort and solution times by orders of magnitude (cf. section 7.1.1).

6.6 Summary and conclusions

In this chapter, the optimization-based conceptual synthesis of a real-world distributed energy supply system is addressed using the automated superstructure-based and the automated superstructure-free synthesis methodologies. A three-step synthesis procedure is presented that generates a set of promising solution alternatives instead of a single optimal solution: First, single-objective synthesis optimization is performed (cf. section 6.2.1): The optimal solution incorporates multiple redundant units and improves the net present value of the base case by 39 %. Secondly, the rich near-optimal solution space of the synthesis problem is explored by generating near-optimal solution alternatives (cf. section 6.3.3). Finally, multi-objective synthesis optimization is performed yielding Pareto-optimal compromise solutions for two conflicting criteria (cf. section 6.4.1): minimization of total investments and minimization of annual cumulative energy demand.

It is shown that both synthesis methodologies automatically identify and evaluate a great many options including conventional, cogeneration, and trigeneration concepts taking into account already existing equipment and constructional limitations. However, only the superstructure-based approach guarantees optimality. Moreover, analysis of the generated solutions provides deeper understanding of the synthesis problem than if only the single optimal solution is considered. In particular, analysis of the near-optimal solution space identifies both common features and differences among the generated solutions. This supports the decision maker to account for the "must haves" of good solutions as well as for practical aspects that have not been explicitly considered during optimization, or that might change in the future, such as flexibility with regard to installation locations and decentralization options, cost for equipment control, flexibility towards changing energy demands, etc. Analysis of the Pareto-optimal compromise solutions enables to evaluate the impact of equipment selection and sizing on the trade-offs between investment cost and CED. For instance, it is shown that to significantly reduce the CED of the NPV-optimal solution, significant higher investments are required for the installation of a third CHP engine. In summary, the proposed framework supports the decision maker to reach rational and far-sighted synthesis decisions through the generation of deep insight into the synthesis problem.

Finally, it should be noted that optimization-based synthesis of distributed energy supply systems using the methodologies proposed in this thesis still requires expert knowledge. However, the required knowledge is limited to energy-related know-how that is usually prevailing among energy experts; in particular, no professional mathematical programming knowledge is required. Thus, the proposed methodologies enable practitioners in the energy sector to perform optimization-based DESS synthesis. Therefore, the superstructure-based and the superstructure-free synthesis methodologies fulfill the requirements for automated optimal synthesis of distributed energy supply systems as discussed in section 2.5 (Table 6.10).

Table 6.10: Comparison of the requirements for an automated synthesis method as discussed in section 2.5 and the features of the synthesis methods proposed in this thesis.

Requirement (cf. section 2.5)	**Feature (☐/☒) (Superstructure-based)**	**Feature (☐/☒) (Superstructure-free)**
• Generic automated synthesis methodology	☒ Successive approach for automated superstructure generation and expansion.	☒ Knowledge-integrated hybrid algorithm, energy conversion hierarchy (ECH).
• Accounting for characteristics of DESS	☒ Algorithm for automated superstructure and model generation, employed MILP formulation.	☒ Energy conversion hierarchy (ECH), employed MILP formulation.
• Near-optimal solutions	☒ Integer-cut approach.	☒ Evolutionary algorithm, no additional computations.
• Multi-objective optimization	☒ ε-constraint method.	☒ Aggregation selection based on the SMS-EMOA.
• Real-world synthesis	☒ Retrofit considering base case equipment and topographic constraints.	☒ Retrofit considering base case equipment and topographic constraints.
• Comparison deterministic/ metaheuristic optimization	☒ See section 6.5.	☒ See section 6.5.

Chapter 7

Summary and conclusions

The conceptual synthesis of distributed energy supply systems (DESS) is an inherently challenging problem that is characterized by time-dependent constraints (e.g., energy demands, ambient temperatures curves, etc.), economy of scale of equipment investment, limited capacities of standardized equipment, and part-load performance characteristics of the considered energy conversion units. Moreover, for optimal DESS synthesis, multiple redundant units are generally to be expected. Optimization-based synthesis methods offer great potentials for the synthesis of cost-effective, energy-efficient, and sustainable systems. However, a lack of adequate, user-friendly methods has so far hindered routine application of optimization in engineering practice.

In research, most commonly, *superstructure-based* synthesis is performed for optimal systems synthesis. In this approach, a user-defined superstructure is analyzed using mathematical programming techniques to identify the optimal solution among the alternatives encoded in the superstructure. Current optimization software facilitates the use of superstructure-based synthesis, e.g., by enabling easy problem definition through graphical superstructure modeling. However, the *a priori* definition of the superstructure remains a serious obstacle for the use of superstructure-based synthesis in industrial practice: On the one hand, the manual superstructure definition bears the risk to exclude the optimum from consideration; on the other hand, the use of excessively large superstructures causes prohibitively large computational effort. To circumvent these drawbacks, *superstructure-generation* methods and *superstructure-free* synthesis methods have been proposed. Superstructure-generation methods automatically define a superstructure for a given synthesis problem. Superstructure-free methods avoid the use of a superstructure by enabling simultaneous alternatives generation and optimization. Available approaches involve several drawbacks that impede their use for the optimal synthesis of distributed energy supply systems: Superstructure-generation methods neglect major DESS characteristics; superstructure-free methods require the manual definition of many technology-specific replacement rules, which is equally difficult as the definition of an appropriate superstructure.

In this thesis, two novel synthesis methodologies are proposed to facilitate the use of optimization for efficient and reliable DESS synthesis, thus making optimization accessible for practitioners: The *automated superstructure-based* and the *superstructure-free* synthesis methodology. The proposed methodologies avoid both the *a priori* definition of a superstructure and the manual definition of many technology-specific replacement rules while accounting for the major characteristics inherent to DESS synthesis problems. The superstructure-based framework (chapter 4) relies on an algorithm for automated superstructure-generation. The method employs a successive superstructure expansion and optimization strategy that continuously increases the number of units included in the superstructure until the optimal solution is identified. The superstructure-free approach (chapter 5) combines evolutionary optimization and deterministic optimization for simultaneous alternatives generation and optimization. A knowledge-integrated mutation operator is proposed that relies on a hierarchically-structured graph, the so-called *energy conversion hierarchy* (ECH). The ECH efficiently defines all reasonable replacement rules, thus avoiding their manual definition. The mutation operator performs structural replacements for the evolutionary generation of solution alternatives. Both synthesis methodologies use a generic component-based modeling framework, thus making the methodologies independent of the employed mathematical programming formulation. In this thesis, a robust MILP formulation is used that allows to simultaneously optimize the structure, sizing, and operation of distributed energy supply systems accounting for time-varying load profiles, continuous equipment sizing, economy of scale of equipment investment, and part-load equipment performance (chapter 3).

For optimization-based retrofit synthesis of a real-world industrial site, an automated three-step synthesis procedure is presented that employs the proposed synthesis methodologies to demonstrate a whole range of features, thus highlighting the practicality of the proposed methodologies (chapter 6). The proposed three-step synthesis procedure generates a set of promising solution alternatives instead of a single optimal solution:

First, both the superstructure-based and the superstructure-free synthesis methodologies automatically and efficiently identify the optimal solution with respect to the net present value (NPV). The identified solution represents a complex trigeneration system incorporating multiple redundant units. Compared to the base case solution, the NPV is improved by 39 %. To identify this solution, both approaches automatically generate and evaluate a great many options including conventional, cogeneration, and trigeneration concepts taking into account already existing equipment and constructional limitations. However, only the superstructure-based approach guarantees optimality.

Second, the proposed methodologies generate a set of promising near-optimal solution alternatives. For this purpose, the superstructure-based approach implements the *integer-cut approach*, which sequentially adds constraints to the problem formulation to exclude already identified solution structures to generate the next best solutions through repetitive optimization. To limit computational effort, in this thesis, the integer-cut approach is not employed together with the successive superstructure expansion strategy, but it is limited to the final superstructure model of the successive approach. In contrast, the superstructure-free approach does not require any additional computations, since it generates alternative solutions during the search for the optimal solution anyway. The considered synthesis problem is characterized by its rich near-optimal solution space: The ten best identified solution structures lie within a range of 0.17 % with respect to the NPV. It is shown that the superstructure-free approach generates a larger variety of structurally different solutions than the superstructure-based approach, however, not all of the best solution alternatives.

Third, multi-objective synthesis concludes the proposed three-step synthesis procedure by generating Pareto-optimal compromise solutions. For this purpose, the presented superstructure-based approach implements the ε-constraint method performing a predefined number of successive optimization runs to generate the Pareto-front. The superstructure-free approach uses the SMS-EMOA, an aggregation selection-based evolutionary algorithm particularly developed for multi-objective optimization. For the real-world synthesis problem, a Pareto-front is generated that enables the decision maker to evaluate the trade-offs between total investments and cumulative energy demand (CED). Moreover, the Pareto-front provides insight into the dependencies of these trade-offs on equipment selection and sizing. It is shown that the NPV-optimal solution lies in a region of the Pareto-front, where moderate additional expenditures only slightly improve the system's energy efficiency. In fact, for significant CED-savings, significant higher investments are required for the installation of a third CHP engine.

The solution set generated by the automated three-step synthesis procedure provides deeper understanding of the synthesis problem than if only the single optimal solution is considered. Analysis of the near-optimal solution space identifies common features and differences among the generated solutions. Analysis of the Pareto-optimal compromise solutions enables to evaluate the impact of equipment selection and sizing on the trade-offs between investment cost and CED. Thus, the automated three-step synthesis procedure opens up wide space for rational and far-sighted synthesis decisions accounting for the "must haves" of good solutions as well as for practical aspects that have not been explicitly considered during optimization.

In summary, it is shown that both synthesis methodologies proposed in this thesis enable practitioners to perform optimization-based synthesis of distributed energy supply systems. It should be pointed out that the use of the proposed synthesis methodologies only requires energy-related expert knowledge that is usually prevailing among engineers active in the field of energy systems synthesis. In particular, no expert knowledge on mathematical programming is required.

Finally, this thesis provides the foundation for future research as discussed in the next section. Last but not least, based on the experience gained during the work on this thesis, the author comments on the necessity of optimization for the conceptual DESS synthesis.

7.1 Future perspective

In the following, recommendations are given for future research. The suggested research studies are prioritized according to the author's opinion regarding

- the uncertainty of the studies' benefits for the methodological framework, i.e., the potential to enable more efficient problem solving or to broaden the range of application of the synthesis methodologies proposed in this thesis; and
- the expected effort to conduct the suggested studies (including the effort of possibly required preliminary research, e.g., to enable more efficient problem solving as required when further model details are considered).

The assigned priorities range from ① to ③ with ① being the highest and ③ being the lowest priority (Fig. 7.1). Note that the prioritization should be understood as recommendation rather than as precise and unambiguous rating. Further note that the structure of this section does not reflect the prioritization of the suggested research topics. Instead, this section is structured according to the classification of the research studies into studies concerning algorithm design and efficient problem formulation, and studies concerning the modeling framework.

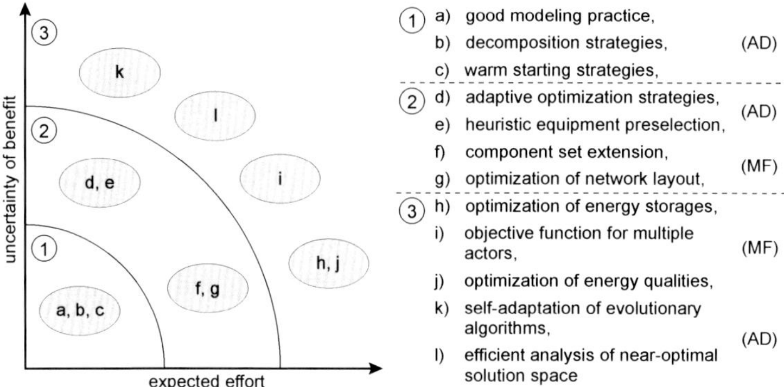

Figure 7.1: Prioritization of the suggested studies for future research: Uncertainty of the studies' benefits plotted against the expected effort to conduct the suggested studies. The studies are classified according to whether they concern algorithm design and problem formulation (AD), or the modeling framework (MF).

7.1.1 Future research on algorithm design and efficient problem formulation

The optimization-based synthesis of the real-world problem considered in this thesis is computationally involved (cf. chapter 6). Thus, the computational performance of the proposed synthesis methods is expected to be a serious obstacle for their successful implementation in engineering practice if more complex synthesis problems are addressed and if further model details are considered (e.g., network layout, energy storages, higher resolution time series, etc.). For this reason, in the following, studies for future research are suggested that aim at the enhancement of both the solution algorithms and the problem formulation to realize more efficient problem solving.

Good modeling practice ①. Good modeling practice concerns sound and efficient problem formulation and algorithm configuration for faster and more robust problem solving. Techniques for good modeling practice take advantage from good use of solver-specific characteristics, such as presolve methods, scaling methods, and branching choices. Moreover, linearization and convexification strategies (e.g., variable separation techniques, piecewise linearization formulations, and substitution techniques) allow to reformulate nonlinear and nonconvex relations of mathematical models to generate good-natured linear or convex nonlinear programming formulations. These reformulated problems are usually much easier to solve, and thus enable more robust and faster optimization-based synthesis. Floudas (1995), Kallrath and Wilson (1997), and Williams (1999) review techniques for good modeling practice for both mixed-integer linear and nonlinear programming problems. The author of this thesis suggest to systematically study techniques for good modeling practice to enhance the synthesis framework proposed in this thesis.

Decomposition methods ①. Sagastizábal (2012) reviews decomposition methods in the context of energy systems optimization. Decomposition methods decompose the original, difficult-to-solve problem into smaller, easy-to-solve subproblems, which are synchronized by a master problem. The master problem usually contains only the easy-to-solve part of the original problem, while the subproblems incorporate the complicating parts of the problem. To guarantee convergence towards the global optimal solution, iterative procedures are used for the synchronization of the master and the subproblems.

Virmani et al. (1989) and Rong et al. (2008) implement the *Lagrangean decomposition* for operation optimization of complex energy systems incorporating heat storages. In these papers, the optimization problem is decomposed by the operational constraints, thus enabling to address each time step in a separate subproblem. In both papers, solution time reductions by orders of magnitude are reported. A detailed review of the Lagrangean decomposition is provided by Guignard (2003).

More recently, Yokoyama and Ose (2012) proposed a decomposition approach specifically tailored to the synthesis of distributed energy supply systems: This approach directly exploits the hierarchical relationship between the upper synthesis and design levels, and the lower operational level of DESS synthesis problems: A bi-level formulation is proposed that is solved by nested optimization algorithms. As with the Lagrangean decomposition, Yokoyama also reports reductions of the solution times by orders of magnitude.

The author of this thesis expects major benefits for the computational performance of the synthesis methods proposed in this thesis if similar decomposition approaches are implemented.

Warm starting strategies ①. Warm starting strategies reduce the computational effort of an optimization computation by employing problem-specific information for good problem initialization (Ralphs and Güzelsoy, 2006). Warm starting strategies are useful, e.g., when a series of closely-related problems are solved. For mixed-integer problems, e.g., the branch-and-bound search tree stores the values of the decision variables defining the current best solution and the values of already calculated bounds. Warm starting methods as presented by Ralphs and Güzelsoy (2006) save the search tree information for problem initialization of further optimization runs.

The successive optimization algorithm of the superstructure-based synthesis method solves a series of structurally similar optimization problems that are stepwise extended by the addition of single units. It is shown that, the further the successive optimization progresses, the smaller are the changes in the optimal solution; however, at the same time, the computational effort increases exponentially with the growing number of considered equipment. Thus, a warm starting strategy is proposed for the successive optimization that initializes the search tree of each next optimization run with the search tree of the preceding optimization run. The author expects significant solution time reductions of the successive optimization when the proposed warm starting strategy is implemented. Note that similarly, the generation of the near-optimal solutions and the Pareto-front generation also solve a series of only gradually modified optimization problems, and thus warm starting is expected to significantly accelerate these techniques as well.

Finally, note that the superstructure-free synthesis method implements a hybrid optimization approach that uses an evolutionary algorithm on the upper level and deterministic MI(N)LP optimization on the lower level. At present, the two optimization algorithms work independently of each other; however, the evolutionary algorithm generates many similar MI(N)LP problems for sizing and operation optimization. Thus, for the solution of the lower level problems, warm starting strategies should be implemented to employ information of preceding evaluations of similar structures.

Adaptive optimization strategies ②. The synthesis methods proposed in this work exploit the special characteristics of distributed energy supply systems for efficient problem solving. To take further advantage of the special DESS characteristics, future research should be conducted on the design of adaptive optimization strategies: The basic concept of the proposed strategies is to gradually approximate the optimal solution of a synthesis problem by using simplified models, which are adaptively refined during the approximation process. This approach converts a complex optimization problem into a set of simpler problems, which can be solved more efficiently while enabling the identification of the optimal solution of the *full model*, i.e., the complex original problem. The proposed adaptation strategies require to solve a series of gradually modified optimization problems, and thus they are inherently suited to make efficient use of warm starting strategies.

The author of this thesis performed preliminary tests on relaxing the lower level MILP operation optimization to a continuous LP problem and on neglecting part-load dependent efficiencies in the equipment models (instead, constant efficiencies are assumed). For both cases, solution time reductions by orders of magnitude are observed. At the same time, optimization of these simplified models yield good bounds for the optimal solution of the full model. Moreover, the approaches yield good starting points for the superstructures of the successive optimization approach. Thus, an adaptation strategy is proposed that employs the results of such simplified models as starting point to gradually approximate both the optimal solution and the final superstructure of the successive approach. In the second step, the solution structure of the simplified model is fixed, and sizing and operation optimization is performed with the full model. From then on, both the optimal solution and the superstructure are refined based on the full model by successively unfixing some of the structural decisions in the superstructure or by adding new units to the superstructure.

Hartono et al. (2012) recently proposed an exergy analysis supported branch-and-bound algorithm that replaces the relaxed NLP subproblems of the traditional branch-and-bound algorithm by exergy analysis calculations. The exergy analysis calculations yield good lower bounds for the available exergy of the current branch at reduced computation times compared to the relaxed NLP computations. However, the proposed algorithm is applicable only to problems with objective functions that can be expressed in terms of exergy. It should be noted, though, that, regardless of the objective function (NPV, energy cost, CED, etc.), optimization-based DESS synthesis generally involves operation optimization, which is related to the identification of exergy-efficient solutions. Thus, the author suggests to study the use of the exergy analysis supported branch-and-bound method in an adaptive optimization strategy such as the one described above, thus enabling to apply the exergy analysis based branch-and-bound method also for DESS synthesis problems with "non-exergetic" objective functions.

Heuristic equipment preselection ②. For conceptual process synthesis, heuristics-based shortcut methods have been proposed for rapid screening of promising process alternatives without the need for detailed specifications. Marquardt et al. (2008) proposed a framework for the optimization-based process design, which combines shortcut methods and rigorous optimization. Inspired by this approach, the author suggests a heuristics-based *equipment preselection* approach to reduce the mathematical complexity of a given synthesis problem by reasonably limiting the set of equipment models regarded during optimization.

Duic et al. (2008) proposed the heuristic *RenewIslands* methodology for the systematic generation of energy supply concepts. As user inputs, the RenewIslands methodology requires only qualitative information about the energy demand levels and the available resources; they are classified as "low", "medium", or "high". Based on these crude and simple user inputs, the RenewIslands methodology uses a range of *if-then*-relations to derive promising configuration alternatives, which are subsequently evaluated by simulation studies (Krajacic et al., 2009).

In future research, an integrated heuristics/optimization-based approach should be studied that employs the RenewIslands methodology as initial step for optimization-based DESS synthesis to preselect the equipment to be considered during optimization. The number of energy conversion units considered for optimization-based DESS synthesis has a strong impact on both the complexity of the mathematical programming problem and the required effort to collect all necessary data for modeling and parameterization of the synthesis problem (Cormio et al., 2003). Thus, the proposed equipment preselection method reduces the complexity of the mathematical programming problem, and it further supports the design engineer to decide which data needs to be collected and refined. In its current state, however, the RenewIslands methodology is tailored to the synthesis of island systems, thus neglecting connections to the public energy market, i.e., the natural gas grid, the electricity grid, etc. Moreover, the RenewIslands method currently neglects already existing equipment. Therefore, the functional range of the RenewIslands methodology needs to be expanded to enable its broad application for equipment preselection to general DESS synthesis problems.

Self-adaptation of evolutionary algorithms ③. One important feature of evolutionary algorithms is that they can use self-adaptation strategies to be tuned "on-the-fly" for the optimization problem at hand. Bartz-Beielstein et al. (2005) proposed the sequential parameter optimization (SPO) technique, a meta-optimization for performance tuning of metaheuristic optimization algorithms. It is shown that SPO can improve the performance of metaheuristic optimization algorithms by more than 90 % with respect to the required number of function evaluations to identify the optimal solution. Moreover, the tuned optimization algorithm is more robust than the default algorithm in the sense that it identifies the optimal solution with higher probability. It is expected that the use of the SPO technique and other tuning strategies will enhance the proposed superstructure-free synthesis methodology. Therefore, in future research, such strategies should be used for the superstructure-free synthesis methodology.

Efficient analysis of near-optimal solution space ③. The results of the real-world problem discussed in this thesis indicate that many near-optimal solutions are to be expected for DESS synthesis problems of practical size. In particular, if the superstructure-based synthesis approach is employed, the generation of these solutions is computationally involved, thus limiting the number of solution alternatives to be generated. To reduce the computational effort for the generation of multiple solutions, Danna et al. (2007) proposed the *one-tree* algorithm for mixed-integer programming problems. This modified branch-and-bound algorithm saves all solutions within a given optimality gap while searching the branch-and-bound tree. It should be noted that the algorithm is readily available in the commercial solver CPLEX. However, applied to the MILP formulation employed in this thesis, the one-tree algorithm generates an unmanageable number of solution alternatives that differ with respect to equipment selection, sizing, and operation; i.e., thousands (!) of solutions are generated, of which most solutions represent structurally similar, or even identical, solutions that differ only with respect to equipment sizing and operation. However, the design engineer is mainly interested in structurally different solutions. For this reason, future studies should be conducted on the use of algorithmic methods that enable automatic selection of structurally different solutions among the generated set of near-optimal solution alternatives. For this purpose, diversity measures and diversity recognition algorithms have been proposed, mostly in the field of evolutionary computation (Manner et al., 1992; Bedau and Zwick, 1995; Morrison and Jong, 2002; Burke et al., 2002). More recently, Danna and Woodruff (2009) proposed algorithmic methods that rely on diversity measures for the selection of a small subset of solutions that maximizes solution diversity from a large solution set. The author assumes that the combination of this selection algorithm with the one-tree algorithm bears the potential to efficiently generate a set of highly diverse near-optimal solution structures using the superstructure-based synthesis method. Thus, this approach should be pursued in future research.

7.1.2 Future research on the modeling framework

Different synthesis problems are dominated by different problem characteristics, and thus different models must be used to adequately reflect the major characteristics of these synthesis problems. Moreover, the modeling decisions are multi-layered and mutually dependent on each other: For instance, if polygeneration units like CHP engines are considered, the underlining time series must reflect the *synchronicity* of important points in time, while the *chronology* of these time steps can be neglected, thus allowing for aggregation of time steps to daily-, weekly-, monthly-averaged time steps. On the other hand, to sufficiently model the operating behavior of a storage system, the underlining time series must reflect the *chronology* of the time course on a finer time scale.

A "complete" DESS modeling framework - if indeed such a thing is possible - enables to account for all model details of any particular DESS synthesis task. However, in this thesis, it is shown that the computational effort to solve real-world problems can become prohibitive already when major model details are neglected, e.g., energy storages. Moreover, some characteristics will dominate others that can thus be neglected without significantly changing the outcome of the optimization. Therefore, it is not reasonable to consider all thinkable model details simultaneously. Instead, the designer needs to balance model accuracy and computational effort when choosing a model paradigm for the synthesis problem at hand. To facilitate the selection of an adequate modeling paradigm for a particular synthesis problem, systematic decision support is desired. To the author's knowledge, there is no such work available in the literature. However, such a framework may be the one real contribution of another Ph.D. thesis. Thus, such a framework is beyond the scope of the following discussion.

In this section, the author gives suggestions for future research that aims at the extension of the current modeling framework by adding model details that have been neglected in this thesis, but which should generally be available to broaden the range of application of the employed modeling framework. It should be emphasized once more that the addition of further model details generally complicates the synthesis problem, thus requiring preliminary research for more efficient problem solving. For this reason, the research topics related to the modeling framework are assigned priorities ② and ③ only.

Component set extension ②. To extend the range of application of the MILP formulation used in this thesis, the set of equipment models needs to be extended to incorporate further units, such as steam and gas turbines (Luo et al., 2012), heat pumps (Juul and Meibom, 2011), solar thermal and photovoltaic panels (Buoro et al., 2012), Organic Rankine Cycles (Sun and Li, 2011), fuel cells (Varbanov et al., 2011), wind turbines (Lund, 2005), etc. The extension of the component set is straightforward. However, the addition of further equipment models is expected to significantly complicate the mathematical programming problems to be solved for DESS synthesis, thus requiring preliminary enhancements of the current synthesis methodologies.

Optimization of network layout ②. Mehleri et al. (2012) show that the network layout strongly impacts both investments (for network installation) and energy cost (due to energy transportation losses along the network), and thus it also influences the optimal solution. Therefore, the MILP modeling framework employed in this work should be extended to integrate the network layout. The extension of the MILP framework is straightforward (Weber and Shah, 2011; Mehleri et al., 2012). However, the integration of the network layout in the synthesis is expected to significantly increase the complexity of the optimization problem. Accordingly, Weber and Shah (2011) and Mehleri et al. (2012) simplify the synthesis problem by neglecting part-load performance of the energy conversion equipment. Thus, future research is required to adequately extend the current modeling framework while still enabling optimization-based DESS synthesis at, for practical applications, reasonable computational cost.

Optimization of energy storages ③. Energy storages strongly influence the optimal synthesis of energy systems (Semadeni, 2004; Dincer, 2004; Ribeiro et al., 2001), in particular through the reduction of required investments, increased total supply capacities, increased system efficiency, and potentials to time-shift energy loads, thus enabling to reduce peak-loads (Hittinger et al., 2012). Moreover, energy storages are key technologies for the employment of renewable energy sources (Kaldellis and Zafirakis, 2007). For these reasons, it is desirable to incorporate energy storage models into the optimization-based synthesis of distributed energy supply systems.

Basically, the incorporation of energy storage models into DESS optimization problems is straightforward (as shown by Söderman and Pettersson (2006) for synthesis optimization and by Blarke and Dotzauer (2011) for operation optimization). However, the consideration of energy storages significantly complicates the optimization problem: First, the consideration of energy storages couples the operational decisions of different time steps, thus translating the quasi-stationary optimization problem into a dynamic optimization problem. Second, additional degrees of freedom are introduced in each time step representing the decisions for storage (dis-)charging. And third, to adequately model the operational behavior of an energy storage system, the time course of the underlying time series generally needs to be captured over the whole period under consideration with an appropriately fine time resolution.

To limit the complexity of the mathematical programming problems including energy storages, Söderman and Pettersson (2006) and Blarke and Dotzauer (2011) assume constant equipment efficiencies. For the same reason, Gupta et al. (2011) assume predefined control strategies for the (dis-)charging of energy storages in their operation optimization formulation. Moreover, for synthesis optimization, Söderman and

Pettersson (2006) vastly reduce the number of time steps considered during optimization to only eight time steps (two time steps, day and night, for each season). Ortiga et al. (2011) use the *typical days* approach to systematically reduce the number of time steps considered during optimization. This approach is based on the assumption that a small set of typical days can be used to adequately represent the time course of a much longer time period (Lozano et al., 2009; Domínguez-Muñoz et al., 2011).

In literature, there is no consent on which level of detail of the system modeling is required to adequately reflect the behavior of energy storage systems while reasonably reducing the problem complexity: Zhou et al. (2013), e.g., conclude that part-load performance characteristics of energy conversion equipment can be neglected if energy storages are considered. However, their conclusion is based on the assumption that only equally-sized equipment can be installed, thus neglecting economy of scale for equipment investments. In summary, the consideration of energy storages for optimization-based DESS synthesis is a challenging task that requires further research to trade-off modeling quality (concerning both equipment modeling and time series modeling) and computational effort.

Objective function for multiple actors ③. For the real-world synthesis problem discussed in this thesis, one company operates both the production processes inducing the energy demands and the energy conversion units meeting these demands. However, distributed energy supply systems generally integrate energy demands and energy conversion units operated by different actors, i.e., energy supply companies, production companies, occupiers of buildings, etc. This situation complicates the synthesis task because an overall optimal solution, e.g., with respect to the net present value generally does not represent an optimal solution for every actor as shown by Jennings et al. (2012). For supply chain optimization, Gjerdrum et al. (2001) proposed a mathematical programming formulation for *optimized profit distribution* to account for multiple enterprises integrated in one supply chain optimization problem: This approach aims at determining an optimal schedule and *transfer prices*, i.e., prices for products transferred between single enterprises, which maximize the profitability for each enterprise. In future research, a similar concept should be developed for distributed energy systems incorporating multiple actors.

Optimization of energy qualities ③. In the modeling framework employed in this work, the quality levels of the energy carriers, i.e., temperatures and pressures, are not subject to optimization. However, the quality levels generally influence the system performance, e.g., the COP of a chiller decreases with decreasing supply and increasing return temperatures. Yet, the inclusion of the quality levels in the optimization generally leads to difficult-to-solve MINLP problems. Thus, approaches in literature introduce simplifying assumptions to limit the overall problem complexity when energy qualities are included in the optimization: Luo et al. (2012), e.g., propose an MINLP model for the operation optimization of complex steam turbine systems. To limit the problem complexity, the quality levels of possible steam extractions are not optimized continuously, but a few discrete quality levels are defined *a priori*. Tveit et al. (2009) incorporate the temperature levels of the heat carriers in the optimization of a district heating network, but they neglect the hydraulics that generally play a major role for these systems. Manassaldi et al. (2011) perform optimal synthesis of a steam heat recovery system, but they regard only a stationary model during a single period. Chen et al. (2011) neglect part-load dependent operation performance. To consider the quality levels in the optimization without oversimplifying the remaining model, Varbanov et al. (2004) and Velasco-Garcia et al. (2011) use decomposition approaches, such as the *Successive*-MILP (S-MILP) approach: In this approach, the values of the heat carrier temperatures are successively changed on the upper *master* level, while they are assumed constant on the lower *slave* level for MILP sizing and operation optimization. In future research, the impact of considering the quality levels of the energy carriers in DESS synthesis problems should be analyzed in detail as should be the adequate problem formulation.

Note that, when energy qualities are included in the optimization, the differences between parallel and serial connections must be reflected by the synthesis methodologies. The superstructure-free approach can readily account for different interconnections by employing the full energy conversion hierarchy. In contrast, the superstructure-based approach requires revision with regard to possible interconnections encoded in the generated superstructures. If the energy qualities are subject to optimization, the superstructures are complicated by serial connections and mixers. Thus, the successive approach must be extended to reflect these substructures in the superstructure expansion procedure.

7.2 A final comment on the necessity of optimization-based synthesis methods

In this thesis, two methodological frameworks are proposed for the automated optimization-based synthesis of distributed energy supply systems on the conceptual level. For the real-world problem considered in this thesis, both methods identify the optimal solution and many structurally different, near-optimal solution alternatives. In fact, the near-optimal solutions differ only marginally with respect to the objective function, and thus are practically equally good. Therefore, two questions arise: 1) Do practitioners actually require optimization-based synthesis methods, or can they perform the synthesis manually along the lines of "*That would be really a stroke of bad luck, if I didn't find one of those many solution structures on my own!*"? 2) Does the design engineer really need to identify the optimal solution, or does any near-optimal solution suffice as starting point for the final decision?

First of all, it is the author's firm belief that a design engineer will practically not be able to identify solutions as those presented in this thesis without the use of optimization-based synthesis methods. To understand this, keep in mind that the solutions do not only differ with respect to the selected equipment, but moreover they strongly differ with respect to equipment sizing and operation. So, even if the design engineer can guess one of the many near-optimal solution structures, the optimal sizing and operation still remain complex tasks, which should be addressed using optimization. The proposed optimization-based synthesis methods integrate the decisions on the three levels *synthesis*, *design*, and *operation*, and therefore are superior to any heuristic approach.

Second, the optimal solution of the mathematical model is generally not the optimal real-world solution, i.e., the solution that will be finally implemented. Thus, the identification of the optimal solution of the mathematical model should not be top priority at any cost, i.e., computational cost. However, the same is true for any other near-optimal solution. Hence, as exhaustively discussed in this thesis, the generation of a set of different solution alternatives is equally important, if not even more important than the generation of a single optimal solution. However, the generation of near-optimal solution alternatives also calls for integrated, automated alternatives generation and evaluation, i.e., optimization. Moreover, the generation of the optimal solution is not necessarily more complex than the identification of the next best alternatives. In fact, for the considered real-world problem, and independently of whether the superstructure-based or the superstructure-free approach is employed, the computational effort to generate the optimal solution is on average the same as for the generation of any of the top 10 solutions. Thus, the optimal solution should preferably be among the generated solutions.

In summary, the author concludes that the design of an optimization-based synthesis method should always aim at enabling to identify the optimal solution. However, for practical problems, only the provision of a set of promising candidate solutions opens up a wide range of rational decision options as desired during the conceptual design phase, thus enabling the design engineer to reach a final decision.

To make a long story short,

'The single best is the enemy of the many good.'
(freely adapted from Voltaire)

Appendix A

Equipment models

A.1 Nonlinear equipment models

A.1.1 Investment cost curves

Investment costs are calculated by capacity power laws (Smith, 2005):

$$I_{\mathrm{N}} = I_{\mathrm{B}} \cdot \left(\frac{\dot{Q}_{\mathrm{N}}}{\dot{Q}_{\mathrm{B}}} \right)^{M},$$

The coefficients $\dot{Q}_{\mathrm{B}}$, I_{B}, and M are listed in table A.1.

Table A.1: Coefficients of capacity power laws for calculation of equipment investments.

technology	$\dot{Q}_{\mathrm{B}}$ / kW	I_{B} / €	M / -
Boiler	100	21 480	0.4502
CHP engine	500	226 397	0.6725
Absorption chiller	50	39 722	0.5002
Turbo-chiller	400	84 444	0.8635

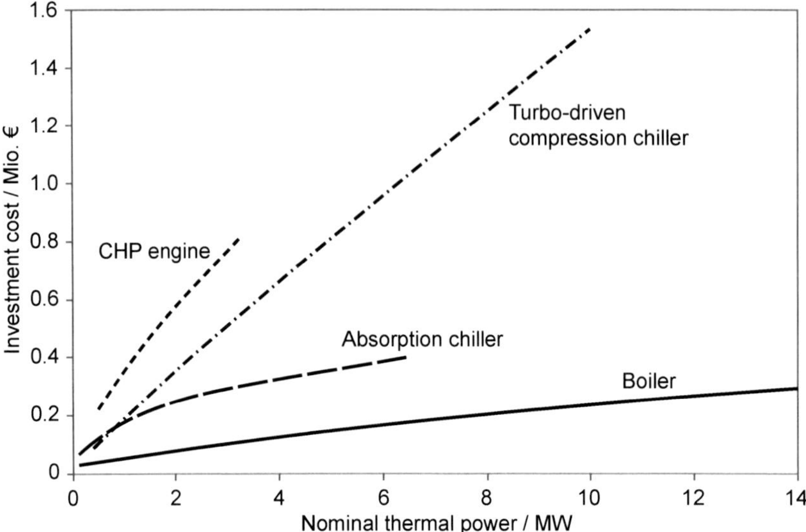

Figure A.1: Investment cost curves of the modeled energy conversion technologies within available capacity ranges.

A.1.2 Part-load performance curves

The characteristic part-load performance curves shown in Fig. A.2 are given by:

- Boiler:

$$\eta/\eta_{\mathrm{N}} = \frac{C_1 + C_2 \cdot q + C_3 \cdot q^2 + C_4 \cdot q^3}{C_5 + C_6 \cdot q + C_7 \cdot q^2 + C_8 \cdot q^3}, \quad q = \frac{\dot{Q}}{\dot{Q}_{\mathrm{N}}},$$

with $C_1 = -0.07557$, $C_2 = 1.39731$, $C_3 = -7.0013$, $C_4 = 21.75378$, $C_5 = 0.03487$, $C_6 = 0.67774$, $C_7 = -5.34196$, and $C_8 = 20.66646$

- CHP engine:

$$\eta/\eta_{\mathrm{N}} = -0.2434 \cdot q^2 + 1.1856 \cdot q + 0.0487, \quad q = \frac{\dot{Q}}{\dot{Q}_{\mathrm{N}}}$$

- Absorption chiller:

$$\mathrm{COP}/\mathrm{COP}_{\mathrm{N}} = \frac{q_0}{C_1 + C_2 \cdot q_0 + C_3 \cdot {q_0}^2}, \quad q_0 = \frac{\dot{Q}_0}{\dot{Q}_{0,\mathrm{N}}},$$

with $C_1 = 0.24999$, $C_2 = -0.0833$, $C_3 = 0.8333$

- Turbo-chiller:

$$\mathrm{COP}/\mathrm{COP}_{\mathrm{N}} = 0.8615 \cdot {q_0}^3 - 3.5494 \cdot {q_0}^2 + 3.679 \cdot q_0 + 0.0126, \quad q_0 = \frac{\dot{Q}_0}{\dot{Q}_{0,\mathrm{N}}}$$

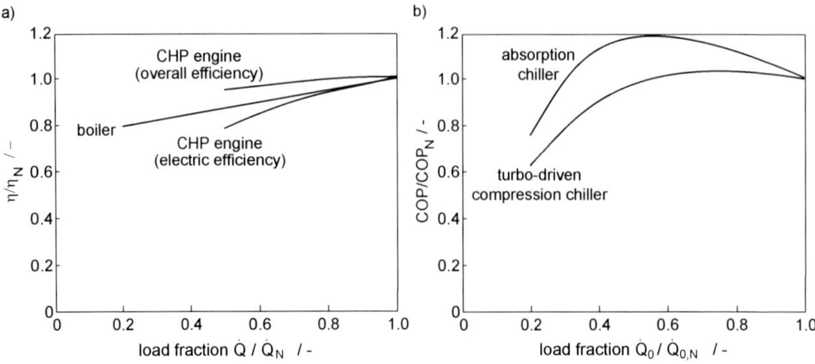

Figure A.2: Characteristic part-load performance curves of heat generators (a) and chillers (b).

A.1.3 Nominal efficiencies of CHP engines

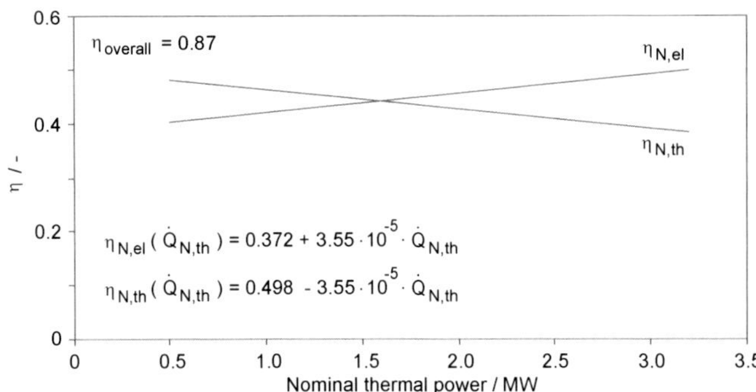

Figure A.3: Nominal electric and thermal efficiency curves of CHP engines. The nominal overall efficiency is constantly 0.87.

A.2 Linearized equipment models

A.2.1 Investment cost curves

For the piecewise linearized investment cost curves as employed in the MILP formulation (cf. section 3.2.1), the nodes of the linear regression models are given in table A.2.

Table A.2: Nodes of piecewise linearized investment cost curves (AC: absorption chiller, TC: turbo-driven compression chiller).

technology	node 1		node 2		node 3	
	$Q_{N,1}$ / kW	I_1 / €	$Q_{N,2}$ / kW	I_2 / €	$Q_{N,3}$ / kW	I_3 / €
Boiler	100	34 343	14 000	379 580	-	-
CHP engine	500	230 022	712	278 644	3200	850 563
AC	50	68 493	750	154 012	6500	522 651
TC	400	89 006	10 000	1 572 302	-	-

A.2.2 Part-load performance curves

For the piecewise linearized characteristic part-load performance curves as shown in Fig. A.4, the nodes of the linear regression models are given in table A.3.

Table A.3: Nodes of piecewise linearized characteristic part-load performance curves (AC: absorption chiller, TC: turbo-driven compression chiller, v: relative output power (load-fraction), u: relative input power).

technology	node 1		node 2		node 3	
	v_1 / -	u_1 / -	v_2 / -	u_2 / -	v_3 / -	u_3 / -
Boiler	0.2	0.2184	1.0	1.0004	-	-
CHP engine	0.5	0.4790	1.0	0.9815	-	-
AC	0.2	0.2722	0.6	0.4833	1.0	0.9833
TC	0.2	0.3185	0.7	0.5936	1.0	0.9828

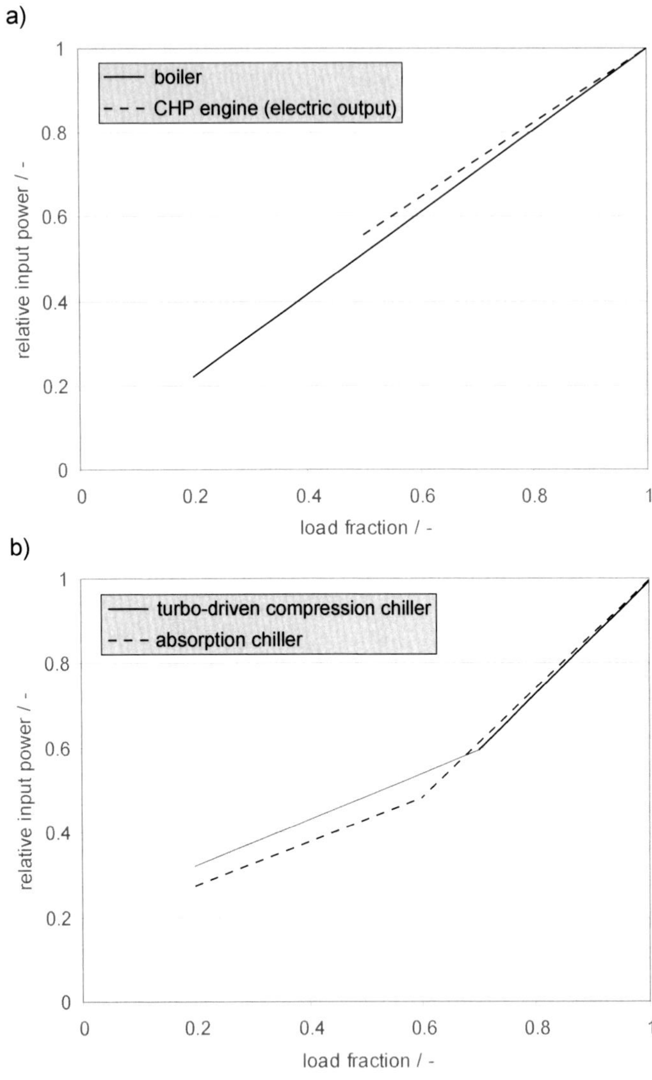

Figure A.4: Linearized performance models of heat generators (a) and chillers (b).

Appendix B

Mutation probabilities

The production probabilities listed in the following sections are calculated by Eqs. (5.6)-(5.14) (cf. section 5.3.3) assuming a size-changing probability of $P_{\Delta\text{Size}} = 1/3$.

B.1 Illustrative example (simplified ECH)

The production probabilities for the illustrative example employing the simplified ECH (Fig. B.1) are given in Table B.1.

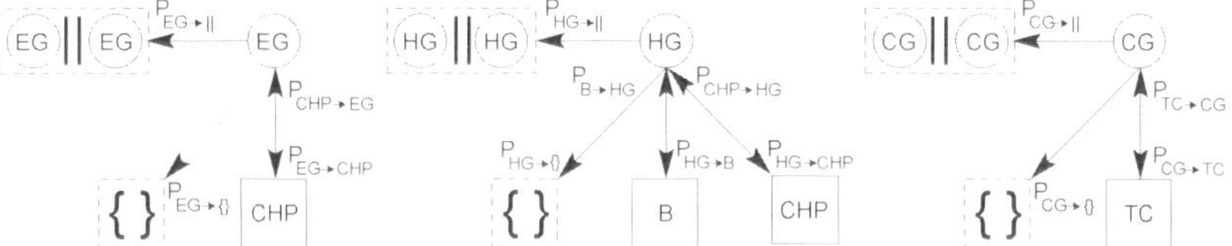

Figure B.1: Tree representation of production set with tagged production probabilities for the simplified ECH.

Table B.1: Probabilities of all productions representing the simplified ECH for the illustrative synthesis problem.

$P_{\text{EG}\to\|\|}$:	1/6	$P_{\text{EG}\to\text{CHP}}$:	2/3	$P_{\text{B}\to\text{HG}}$:	1.0
$P_{\text{EG}\to\{\}}$:	1/6	$P_{\text{CG}\to\text{TC}}$:	2/3	$P_{\text{CHP}\to\text{EG}}$:	1.0
$P_{\text{HG}\to\|\|}$:	1/6	$P_{\text{HG}\to\text{B}}$:	1/3	$P_{\text{CHP}\to\text{HG}}$:	1.0
$P_{\text{HG}\to\{\}}$:	1/6	$P_{\text{HG}\to\text{CHP}}$:	1/3	$P_{\text{TC}\to\text{CG}}$:	1.0
$P_{\text{CG}\to\|\|}$:	1/6				
$P_{\text{CG}\to\{\}}$:	1/6				

B.2 Illustrative example (extended ECH)

The production probabilities for the illustrative example employing the extended ECH (Fig. B.2) are given in Table B.2.

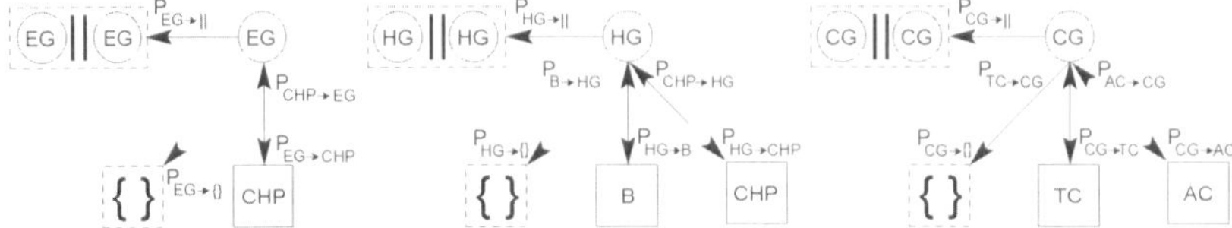

Figure B.2: Tree representation of production set with tagged production probabilities for the extended ECH.

Table B.2: Probabilities of all productions representing the extended ECH for the illustrative synthesis problem.

$P_{EG\to\|\|}$:	1/6	$P_{EG\to CHP}$:	2/3	$P_{B\to HG}$:	1.0
$P_{EG\to\{\}}$:	1/6	$P_{CG\to TC}$:	1/3	$P_{CHP\to EG}$:	1.0
$P_{HG\to\|\|}$:	1/6	$P_{CG\to AC}$:	1/3	$P_{CHP\to HG}$:	1.0
$P_{HG\to\{\}}$:	1/6	$P_{HG\to CHP}$:	1/3	$P_{TC\to CG}$:	1.0
$P_{CG\to\|\|}$:	1/6	$P_{HG\to B}$:	1/3	$P_{AC\to CG}$:	1.0
$P_{CG\to\{\}}$:	1/6				

B.3 Real-world example

The production probabilities for the real-world example employing the extended ECH (Fig. B.3) are given in Table B.3.

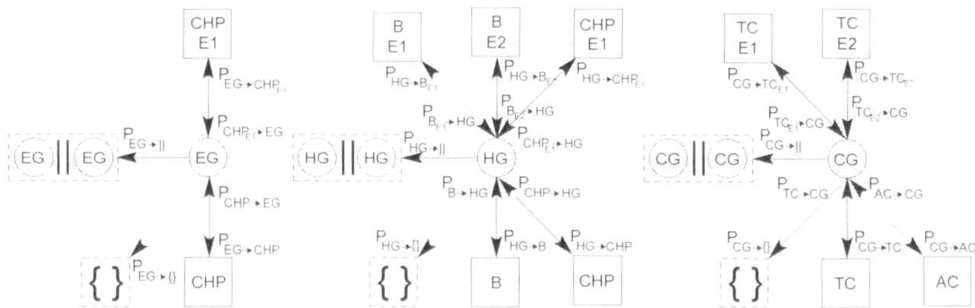

Figure B.3: Tree representation of production set with tagged production probabilities for the real-world synthesis problem.

Table B.3: Probabilities of all productions representing the extended ECH for the real-world synthesis problem.

$P_{\mathrm{EG}\to\|\|}$:	1/6	$P_{\mathrm{EG}\to\mathrm{CHP}}$:	1/3	$P_{\mathrm{CG}\to\mathrm{TC}}$:	1/6	$P_{\mathrm{B}\to\mathrm{HG}}$:	1.0
$P_{\mathrm{EG}\to\{\}}$:	1/6	$P_{\mathrm{EG}\to\mathrm{CHP}_{\mathrm{E1}}}$:	1/3	$P_{\mathrm{CG}\to\mathrm{AC}}$:	1/6	$P_{\mathrm{CHP}\to\mathrm{EG}}$:	1.0
$P_{\mathrm{HG}\to\|\|}$:	1/6	$P_{\mathrm{HG}\to\mathrm{B}}$:	2/15	$P_{\mathrm{CG}\to\mathrm{TC}_{\mathrm{E1}}}$:	1/6	$P_{\mathrm{CHP}\to\mathrm{HG}}$:	1.0
$P_{\mathrm{HG}\to\{\}}$:	1/6	$P_{\mathrm{HG}\to\mathrm{CHP}}$:	2/15	$P_{\mathrm{CG}\to\mathrm{TC}_{\mathrm{E2}}}$:	1/6	$P_{\mathrm{B}_{\mathrm{E1}}\to\mathrm{HG}}$:	1.0
$P_{\mathrm{CG}\to\|\|}$:	1/6	$P_{\mathrm{HG}\to\mathrm{B}_{\mathrm{E1}}}$:	2/15			$P_{\mathrm{B}_{\mathrm{E2}}\to\mathrm{HG}}$:	1.0
$P_{\mathrm{CG}\to\{\}}$:	1/6	$P_{\mathrm{HG}\to\mathrm{B}_{\mathrm{E2}}}$:	2/15			$P_{\mathrm{CHP}_{\mathrm{E1}}\to\mathrm{HG}}$:	1.0
		$P_{\mathrm{HG}\to\mathrm{CHP}_{\mathrm{E1}}}$:	2/15			$P_{\mathrm{TC}\to\mathrm{CG}}$:	1.0
						$P_{\mathrm{AC}\to\mathrm{CG}}$:	1.0
						$P_{\mathrm{TC}_{\mathrm{E1}}\to\mathrm{CG}}$:	1.0
						$P_{\mathrm{TC}_{\mathrm{E2}}\to\mathrm{CG}}$:	1.0

Appendix C

Numerical Results

C.1 Evaluation of MILP formulation for DESS synthesis problems

Table C.1: Optimal solution of MILP test problem employing a linearized model with three CHP partitions and at maximum three nodes for each piecewise linearization (cf. appendix A.2): Nominal thermal powers, overall efficiencies, COPs, investment cost, operating times and annual average part-loads of the installed equipment (B: boiler, CHP: CHP engine, TC: turbo-chiller, AC: absorption chiller).

	B1	B2	CHP1	TC1	TC2	AC1
Thermal power / kW	1900	100	2300	1900	843	367
Investment / k€	79	34	600	321	157	107
η_N ($\eta_{N,th}$, $\eta_{N,el}$), COP_N / -	0.9	0.9	0.87 (0.434, 0.435)	5.54	5.54	0.67
operating time / h/a	< 100	2190	8760	8760	4380	2190
average part-load / %	100	100	70	72	70	94

Table C.2: Optimal solution of MINLP test problem: Nominal thermal powers, overall efficiencies, COPs, investment cost, operating times and annual average part-loads of the installed equipment (B: boiler, CHP: CHP engine, TC: turbo-chiller, AC: absorption chiller).

	B1	B2	CHP1	TC1	TC2	AC1
Thermal power / kW	1900	284	2116	1900	798	412
Investment / k€	80	36	579	322	151	121
η_{N} ($\eta_{\mathrm{N,th}}$, $\eta_{\mathrm{N,el}}$), $\mathrm{COP}_{\mathrm{N}}$ / -	0.9	0.9	0.87 (0.424, 0.447)	5.54	5.54	0.67
operating time / h/a	$<$ 100	2190	8760	8760	4380	2200
average part-load / %	100	100	73	73	72	75

C.2 Illustrative example of automated superstructure-based synthesis

Table C.3: Optimal solution of illustrative grassroots synthesis problem: Nominal thermal powers, overall efficiencies, COPs, investment cost, operating times and annual average part-loads of the installed equipment (B: boiler, CHP: CHP engine, TC: turbo-chiller, AC: absorption chiller).

	B1	B2	CHP1	TC1	TC2	AC1
Thermal power / kW	1900	100	2300	1900	843	367
Investment / k€	79	34	600	321	157	107
η_N ($\eta_{N,th}$, $\eta_{N,el}$), COP_N / -	0.9	0.9	0.87 (0.434, 0.435)	5.54	5.54	0.67
operating time / h/a	< 100	2190	8760	8760	4380	2190
average part-load / %	100	100	70	72	70	94

C.3 Illustrative example of superstructure-free synthesis

C.3.1 Simplified energy conversion hierarchy

In optimal configuration (Table C.4), the heating system consists of one boiler (boiler 1: 1.9 MW) and two CHP engines (CHP engine 1: 1.5 MW, CHP engine 2: 0.9 MW). For cooling, two compression chillers (turbo-chiller 1: 1.9 MW, turbo-chiller 2: 1.2 MW) are installed. The boiler is reserved to solely meet heating peak-loads during summer. CHP engine 1 is operated during winter, spring, and fall. CHP engine 2 is operated during winter and summer, only. Turbo-chiller 1 is operated year-round, while turbo-chiller 2 is operated only during fall and summer. A considerable amount of the electricity produced by the CHP engines can be used on-site to drive the compression chillers.

Table C.4: Optimal solution of illustrative grassroots synthesis problem (simplified ECH): Nominal thermal powers, overall efficiencies, COPs, investment cost, operating times and annual average part-loads of the installed equipment (B: boiler, CHP: CHP engine, TC: turbo-chiller).

	B1	CHP1	CHP2	TC1	TC2
Thermal power / kW	1900	1500	900	1900	1210
Investment / k€	79	498	343	321	214
η_N ($\eta_{N,th}$, $\eta_{N,el}$), COP_N / -	0.9	0.87 (0.448, 0.422)	0.87 (0.467, 0.403)	5.54	5.54
operating time / h/a	< 100	6570	4380	8760	4380
average part-load / %	100	100	86	70	70

Equipment sizing allows for load sharing enabling full-load operation of the boiler and CHP engine 1 whenever operated (Fig. C.1). CHP engine 2 is run at high loads as well (on average at 86 % part-load). The turbo-chillers are always operated at maximum COPs (average part-load of 70 %).

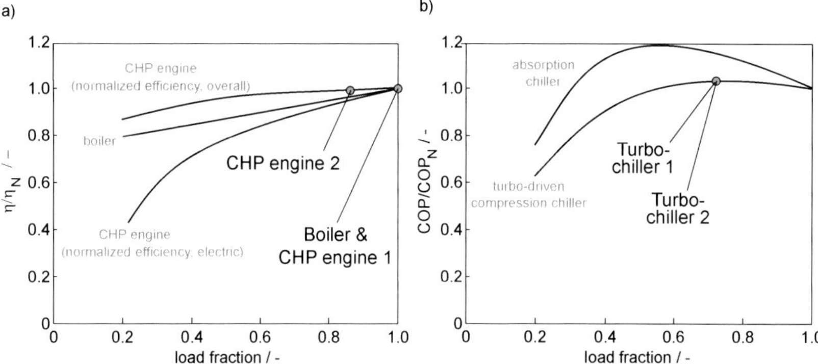

Figure C.1: Annual average operating part-loads of equipment installed in optimal grassroots solution (simplified energy conversion hierarchy).

C.3.2 Extended energy conversion hierarchy

When the superstructure-free synthesis approach is employed with the extended energy conversion hierarchy, the best solution found equals the optimal solution found when the automated superstructure-based synthesis approach is employed (for details, see section 4.3 and appendix C.2).

C.4 Real world example

C.4.1 NPV-optimal solution

Table C.5: Optimal retrofit solution: Nominal thermal powers, overall efficiencies, investment cost, operating times and annual average part-loads of the installed heat generators (E: Existing equipment. N: New equipment. B: boiler. CHP: CHP engine.).

	B E1	CHP N1	CHP N2
Thermal power / kW	7000	2410	2410
Investment / k€	0	668	668
η_N, COP_N / -	0.8	0.87	0.87
operating time / h/a	< 100	8760	8760
average part-load / %	43	100	100

Table C.6: Optimal retrofit solution: Nominal thermal powers, overall efficiencies, COPs, investment cost, operating times and annual average part-loads of the installed chillers (E: Existing equipment. N: New equipment. AC: absorption chiller. TC: turbo-chiller.).

	Site A				Site B		
	AC N1	TC E1	TC N1	TC N2	AC N2	TC N3	TC N4
Thermal power / kW	2610	8000	1090	970	630	670	430
Investment / k€	280	0	196	177	156	131	94
η_N, COP_N / -	0.67	2.8	5.54	5.54	0.67	5.54	5.54
operating time / h/a	5840	< 100	6570	7300	5110	5840	6570
average part-load / %	57	95	81	77	65	71	77

Bibliography

Aguilar, O., Perry, S., Kim, J.-K., and Smith, R. (2007a). Design and optimization of flexible utility systems subject to variable conditions: Part 1: Modelling framework. *Chem. Eng. Res. Des.*, 85(8):1136 – 1148.

Aguilar, O., Perry, S., Kim, J.-K., and Smith, R. (2007b). Design and optimization of flexible utility systems subject to variable conditions: Part 2: Methodology and applications. *Chem. Eng. Res. Des.*, 85(8):1149 – 1168.

Alarcon-Rodriguez, A., Ault, G., and Galloway, S. (2010). Multi-objective planning of distributed energy resources: A review of the state-of-the-art. *Renew. Sustain. Energ. Rev.*, 14(5):1353 – 1366.

Andrecovich, M. and Westerberg, A. (1985). An MILP formulation for heat integrated distillation sequence synthesis. *AIChE J.*, 31(9):1461 – 1474.

Androulakis, I. and Venkatasubramanian, V. (1991). A genetic algorithmic framework for process design and optimization. *Comput. Chem. Eng.*, 15(4):217 – 228.

Angelov, P. P., Zhang, Y., Wright, J. A., Hanby, V. I., and Buswell, R. A. (2003). Automatic design synthesis and optimization of component-based systems by evolutionary algorithms. In Cantú-Paz, E. and Foster, J. A., editors, *GECCO '03: Proceedings of the Genetic and Evolutionary Computation Conference*, volume 2724 of *Lecture Notes in Computer Science*, pages 1938 – 1950, Chicago, Illinois. Springer-Verlag.

Augenstein, E., Herbergs, S., Kuperjans, I., and Lucas, K. (2005). Simulation of industrial energy supply systems with integrated cost optimization. In Kjelstrup, S. and Hustad, J. E., editors, *Proceedings of ECOS 2005: 18th International Conference on Efficiency, Cost, Optimization, Simulation and Environmental Impact of Energy Systems*, pages 627 – 634, Trondheim, Norway. Tapir Academic Press.

Avramenko, Y., Kraslawski, A., and Krysiak, W. (2005). Case-based reasoning tool for the support of process and product design. In Puigjaner, L. and Espuña, A., editors, *European Symposium on Computer-Aided Process Engineering-15, 38th European Symposium of the Working Party on Computer Aided Process Engineering*, volume 20 of *Computer Aided Chemical Engineering*, pages 697 – 702. Elsevier.

Bakshi, B. R. and Fiksel, J. (2003). The quest for sustainability: Challenges for process systems engineering. *AIChE J.*, 49(6):1350 – 1358.

Balas, E. and Jeroslow, R. (1972). Canonical cuts on unit hypercube. *SIAM J. Appl. Math.*, 23(1):61 – 69.

Baños, R., Manzano-Agugliaro, F., Montoya, F., Gil, C., Alcayde, A., and Gómez, J. (2011). Optimization methods applied to renewable and sustainable energy: A review. *Renew. Sustain. Energ. Rev.*, 15(4):1753 – 1766.

Barnicki, S. D. and Siirola, J. J. (2004). Process synthesis prospective. *Comput. Chem. Eng.*, 28(4):441 – 446.

Bartz-Beielstein, T., Lasarczyk, C., and Preuss, M. (2005). Sequential parameter optimization. In McKay, B. and Corne, D., editors, *Proceedings 2005 Congress on Evolutionary Computation (CEC'05)*, volume 1, pages 773 – 780, Edinburgh, Scotland. IEEE Press, Piscataway NJ.

Bedau, M. and Zwick, M. (1995). Variance and uncertainty measures of population diversity dynamics. *Advances in systems science and applications*, 1:7 – 12.

Belton, V. and Stewart, T. J. (2002). *Multiple Criteria Decision Analysis. An integrated approach.* Kluwer Academic Publishers, Boston.

Bertok, B., Kalauz, K., Sule, Z., and Friedler, F. (2013). Combinatorial algorithm for synthesizing redundant structures to increase reliability of supply chains: Application to biodiesel supply. *Ind. Eng. Chem. Res.*, 52(1):181 – 186.

Beume, N., Naujoks, B., and Emmerich, M. (2007). SMS-EMOA: Multiobjective selection based on dominated hypervolume. *Eur. J. Oper. Res.*, 181(3):1653 – 1669.

Beyer, H.-G. (2001). *The theory of evolution strategies.* Springer-Verlag, Berlin/Heidelberg, 1 edition.

Beyer, H. G. and Schwefel, H. P. (2002). Evolution strategies - a comprehensive introduction. *Nat. Comput.*, 1(1):3 – 52.

Biegler, L. T. (2010). *Nonlinear Programming: Concepts, Algorithms, and Applications to Chemical Processes.* MPS-SIAM Series on Optimization. Society for Industrial and Applied Mathematics, Philadelphia.

Biegler, L. T. and Grossmann, I. E. (2004). Retrospective on optimization. *Comput. Chem. Eng.*, 28(8):1169 – 1192.

Biegler, L. T., Grossmann, I. E., and Westerberg, A. W. (1997). *Systematic methods of chemical process design.* Prentice-Hall, Englewood Cliffs, New Jersey.

Blarke, M. B. and Dotzauer, E. (2011). Intermittency-friendly and high-efficiency cogeneration: Operational optimisation of cogeneration with compression heat pump, flue gas heat recovery, and intermediate cold storage. *Energy*, 36(12):6867 – 6878.

Bloch, J. (2008). *Effective Java: A Programming Language Guide (Java Series).* Addison-Wesley Longman, Amsterdam, 2nd edition.

Bonami, P., Biegler, L. T., Conn, A. R., Cornuéjols, G., Grossman, I. E., Laird, C. D., Lee, J., Lodi, A., Margot, F., Sawaya, N., and Wächter, A. (2007). An algorithmic framework for convex mixed integer nonlinear programs. Technical Report Paper 270, Tepper School of Business.

Bondy, J. A. and Murty, U. S. R. (1976). *Graph theory with applications.* Elsevier Science Ltd, Amsterdam.

Book, R. V. (1987). Thue systems as rewriting systems. *J. Symbolic Comput.*, 3(1-2):39 – 68.

Botar-Jid, C. C., Avramenko, Y., Kraslawski, A., and Agachi, P.-S. (2010). Case-based selection of a model of a reverse flow reactor. *Chem. Eng. Process: Process Intensification*, 49(1):74 – 83.

Bouffard, F. and Kirschen, D. S. (2008). Centralised and distributed electricity systems. *Energ. Pol.*, 36(12):4504 – 4508.

Bouvy, C. and Lucas, K. (2007). Multicriterial optimisation of communal energy supply concepts. *Energ. Convers. Manag.*, 48(11):2827 – 2835.

Brooke, A., Kendrick, D., and Meeraus, A. (2010). *GAMS: A user's guide. Tutorial by Rick Rosenthal.* GAMS Development Corporation, Washington, DC, USA.

Brownlee, J. (2011). *Clever Algorithms: nature-inspired programming recipes.* Lulu Enterprises, Raleigh, North Carolina.

Bruno, J., Fernandez, F., Castells, F., and Grossmann, I. (1998). A rigorous MINLP model for the optimal synthesis and operation of utility plants. *Chem. Eng. Res. Des.*, 76(3):246 – 258.

Buoro, D., Casisi, M., Pinamonti, P., and Reini, M. (2011). Optimization of distributed trigeneration systems integrated with heating and cooling micro-grids. *Distr. Generat. Alternative Energ. J.*, 26(2):7 – 34.

Buoro, D., Casisi, M., Pinamonti, P., and Reini, M. (2012). Optimal synthesis and operation of advanced energy supply systems for standard and domotic home. *Energ. Convers. Manag.*, 60(0):96 – 105.

Burke, E. K., Gustafson, S., and Kendall, G. (2002). A survey and analysis of diversity measures in genetic programming. In *GECCO '02: Proceedings of the Genetic and Evolutionary Computation Conference*, GECCO '02, pages 716 – 723, San Francisco, CA, USA. Morgan Kaufmann Publishers Inc.

Caballero, J. A., Odjo, A., and Grossmann, I. E. (2007). Flowsheet optimization with complex cost and size functions using process simulators. *AIChE J.*, 53(9):2351 – 2366.

Casisi, M., Pinamonti, P., and Reini, M. (2009). Optimal lay-out and operation of combined heat and power (CHP) distributed generation systems. *Energy*, 34(12):2175 – 2183.

Cerný, V. (1985). Thermodynamical approach to the traveling salesman problem: An efficient simulation algorithm. *J. Optim. Theor. Appl.*, 45(1):41 – 51.

Chen, C.-L. and Lin, C.-Y. (2011). A flexible structural and operational design of steam systems. *Appl. Therm. Eng.*, 31(13):2084 – 2093.

Chen, Y., Adams, T. A., and Barton, P. I. (2011). Optimal design and operation of flexible energy polygeneration systems. *Ind. Eng. Chem. Res.*, 50(8):4553 – 4566.

Chou, C. C. and Shih, Y. S. (1987). A thermodynamic approach to the design and synthesis of plant utility systems. *Ind. Eng. Chem. Res.*, 26(6):1100 – 1108.

Coello, C. C., Lamont, G., and van Veldhuizen, D. (2007). *Evolutionary Algorithms for Solving Multi-Objective Problems.* Genetic and Evolutionary Computation. Springer-Verlag, Berlin/Heidelberg, 2nd edition.

Connolly, D., Lund, H., Mathiesen, B., and Leahy, M. (2010). A review of computer tools for analysing the integration of renewable energy into various energy systems. *Appl. Energ.*, 87(4):1059 – 1082.

Cormio, C., Dicorato, M., Minoia, A., and Trovato, M. (2003). A regional energy planning methodology including renewable energy sources and environmental constraints. *Renew. Sustain. Energ. Rev.*, 7(2):99 – 130.

Cremer, K., Gruner, S., and Nagl, M. (1999). Handbook of graph grammars and computing by graph transformation. chapter Graph transformation based integration tools: application to chemical process engineering, pages 369 – 394. World Scientific Publishing Co., Inc., River Edge, NJ, USA.

Danna, E., Fenelon, M., Gu, Z., and Wunderling, R. (2007). Generating multiple solutions for mixed integer programming problems. In Fischetti, M. and Williamson, D., editors, *Integer Programming and Combinatorial Optimization*, volume 4513 of *Lecture Notes in Computer Science*, pages 280 – 294. Springer-Verlag, Berlin/Heidelberg.

Danna, E. and Woodruff, D. L. (2009). How to select a small set of diverse solutions to mixed integer programming problems. *Oper. Res. Lett.*, 37(4):255 – 260.

d'Anterroches, L. and Gani, R. (2005). Group contribution based process flowsheet synthesis, design and modelling. *Fluid Phase Equilibria*, 228-229:141 – 146.

Dimopoulos, G. G. and Frangopoulos, C. A. (2008). Optimization of energy systems based on evolutionary and social metaphors. *Energy*, 33(2):171 – 179.

Dincer, I. (2004). Thermal energy storage. In Cleveland, C. J., editor, *Encyclopedia of Energy*, pages 65 – 78. Elsevier, New York.

Doman, L. E. (2011). *International Energy Outlook 2011*, chapter World energy demand and economic outlook, pages 9 – 24. U.S. Energy Information Administration, Washington, DC. Available online. http://www.eia.gov/forecasts/ieo/index.cfm.

Domínguez-Muñoz, F., Cejudo-López, J. M., Carrillo-Andrés, A., and Gallardo-Salazar, M. (2011). Selection of typical demand days for CHP optimization. *Energ. Build.*, 43(11):3036 – 3043.

Douglas, J. M. (1985). A hierarchical decision procedure for process synthesis. *AIChE J.*, 31(3):353 – 362.

Drumm, C., Busch, J., Dietrich, W., Eickmans, J., and Jupke, A. (2013). STRUCTese ® - Energy efficiency management for the process industry. *Chem. Eng. Process: Process Intensification*, 67(0):99 – 110.

Duic, N., Krajacic, G., and da Graça Carvalho, M. (2008). RenewIslands methodology for sustainable energy and resource planning for islands. *Renew. Sustain. Energ. Rev.*, 12(4):1032 – 1062.

Duran, M. and Grossmann, I. (1986). An outer-approximation algorithm for a class of mixed-integer nonlinear programs. *Math. Program.*, 36:307 – 339.

Eberhart and Shi, Y. (2001). Particle swarm optimization: developments, applications and resources. In *Evolutionary Computation 2001*, volume 1, pages 81 – 86.

Edgar, T. F., Himmelblau, D. M., and Lasdon, L. S. (2001). *Optimization of chemical processes*. McGraw-Hill, 2 edition.

Ehrgott, M., Figueira, J., and Greco, S. (2010). *Trends in Multiple Criteria Decision Analysis*. International Series in Operations Research & Management Science. Springer-Verlag, Berlin/Heidelberg.

Eiben, A. E. and Smith, J. E. (2003). *Introduction to evolutionary computing*. Springer-Verlag, Berlin/Heidelberg, 1 edition.

Emmerich, M. (2002). Optimisation of thermal power plant designs: a graph-based adaptive search approach. In Parmee, I. C., editor, *Adaptive computing in design and manufacture V*, pages 87 – 99. Springer-Verlag, Berlin/Heidelberg, 1 edition.

Emmerich, M., Grötzner, M., and Schütz, M. (2001). Design of graph-based evolutionary algorithms: a case study for chemical process networks. *Evol. Comput.*, 9:329 – 354.

Falk, J. E. and Soland, R. M. (1969). An algorithm for separable nonconvex programming problems. *Manag. Sci.*, 15(9):550 – 569.

Fan, L. T., Kim, Y., Yun, C., Park, S. B., Park, S., Bertok, B., and Friedler, F. (2009). Design of optimal and near-optimal enterprise-wide supply networks for multiple products in the process industry. *Ind. Eng. Chem. Res.*, 48(4):2003 – 2008.

Fan, L. T., Lin, Y.-C., Shafie, S., Bertok, B., and Friedler, F. (2012). Exhaustive identification of feasible pathways of the reaction catalyzed by a catalyst with multiactive sites via a highly effective graph-theoretic algorithm: application to ethylene hydrogenation. *Ind. Eng. Chem. Res.*, 51(6):2548 – 2552.

Fan, L. T., Lin, Y.-C., Shafie, S., Hohn, K. L., Bertok, B., and Friedler, F. (2008). Graph-theoretic and energetic exploration of catalytic pathways of the water-gas shift reaction. *J. Chin. Inst. Chem. Eng.*, 39(5):467 – 473.

Farkas, T., Rev, E., and Lelkes, Z. (2005). Process flowsheet superstructures: Structural multiplicity and redundancy: Part I: Basic GDP and MINLP representations. *Comput. Chem. Eng.*, 29(10):2180 – 2197.

Fazlollahi, S., Mandel, P., Becker, G., and Maréchal, F. (2012). Methods for multi-objective investment and operating optimization of complex energy systems. *Energy*, 45(1):12 – 22.

Figueira, J., Greco, S., and Ehrgott, M. (2005). *Multiple Criteria Decision Analysis. State of the Art Surveys.* Springer-Verlag, Berlin/Heidelberg.

Fishbone, L. G. and Abilock, H. (1981). Markal, a linear-programming model for energy systems analysis: Technical description of the bnl version. *Int. J. Energ. Res.*, 5(4):353 – 375.

Floudas, C. A. (1987). Separation synthesis of multicomponent feed streams into multicomponent product streams. *AIChE J.*, 33(4):540 – 550.

Floudas, C. A. (1995). *Nonlinear and Mixed-Integer Optimization: Fundamentals and Applications (Topics in Chemical Engineering).* Oxford University Press, USA.

Floudas, C. A. and Lin, X. (2004). Continuous-time versus discrete-time approaches for scheduling of chemical processes: a review. *Comput. Chem. Eng.*, 28(11):2109 – 2129.

Fraga, E. (1998). The generation and use of partial solutions in process synthesis. *Chem. Eng. Res. Des.*, 76(1):45 – 54.

Fraga, E. S. (2009). A rewriting grammar for heat exchanger network structure evolution with stream splitting. *Eng. Optim.*, 41(9):813 – 831.

Fraga, E. S., Steffens, M. A., Bogle, I. D. L., and Hind, A. K. (2000). An object oriented framework for process synthesis and optimization. In Malone, M. F., Trainham, J. A., and Carnahan, B., editors, *5th International Conference on Chemical Process Design*, volume 323 of *AIChE Symposium Series*, pages 446 – 449.

Frangopoulos, C. A., von Spakovsky, M. R., and Sciubba, E. (2002). A brief review of methods for the design and synthesis optimization of energy systems. *Int. J. Appl. Therm.*, 5(4):151 – 160.

Friedler, F., Tarjan, K., Huang, Y., and Fan, L. (1992). Combinatorial algorithms for process synthesis. *Comput. Chem. Eng.*, 16(Supplement 1):313 – 320.

Friedler, F., Tarjan, K., Huang, Y., and Fan, L. (1993). Graph-theoretic approach to process synthesis: Polynomial algorithm for maximal structure generation. *Comput. Chem. Eng.*, 17(9):929 – 942.

Friedler, F., Varga, J. B., Feher, E., and Fan, L. (1996). Combinatorially accelerated branch-and-bound method for solving the MIP model of process network synthesis. In Floudas, C. and Pardalos, P., editors, *State of the Art in Global Optimization*, pages 609 – 626. Kluwer Academic Publishers, Boston, MA, USA.

Fritsche, U. R. and Schmidt, K. (2007). *Global Emission Model for Integrated Systems (GEMIS)*. Institute of Applied Ecology (Öko-Institut e.V.). online: http://www.gemis.de.

Furman, K. C. and Sahinidis, N. V. (2002). A critical review and annotated bibliography for heat exchanger network synthesis in the 20th century. *Ind. Eng. Chem. Res.*, 41(10):2335 – 2370.

Gebhardt, M., Kohl, H., and Steinrötter, T. (2002). *Preisatlas: Ableitung von Kostenfunktionen für Komponenten der rationellen Energienutzung*. Institute of Energy and Environmental Technology e.V. (IUTA), Duisburg, Germany. Technical Report No.: S 511.

Gill, P., Murray, W., and Saunders, M. (2005). SNOPT: An SQP algorithm for large-scale constrained optimization. *SIAM Review*, 47(1):99 – 131.

Gjerdrum, J., Shah, N., and Papageorgiou, L. G. (2001). Transfer prices for multienterprise supply chain optimization. *Ind. Eng. Chem. Res.*, 40(7):1650 – 1660.

Glover, F. (1975). Improved linear integer programming formulations of nonlinear integer problems. *Manag. Sci.*, 22(4):455 – 460.

Greistorfer, P., Løkketangen, A., Voß, S., and Woodruff, D. (2008). Experiments concerning sequential versus simultaneous maximization of objective function and distance. *J. Heuristics*, 14:613 – 625.

Grimaldi, R. (1998). *Discrete and combinatorial mathematics: An applied introduction (4th edition)*. Addison-Wesley.

Grossmann, I. E. (2012). Advances in mathematical programming models for enterprise-wide optimization. *Comput. Chem. Eng.*, 47:2 – 18.

Grossmann, I. E. and Biegler, L. T. (2004). Part II. Future perspective on optimization. *Comput. Chem. Eng.*, 28(8):1193 – 1218.

Grossmann, I. E. and Guillén-Gosálbez, G. (2010). Scope for the application of mathematical programming techniques in the synthesis and planning of sustainable processes. *Comput. Chem. Eng.*, 34(9):1365 – 1376.

Guignard, M. (2003). Lagrangean relaxation. *Top*, 11:151 – 200.

Guillen-Gosalbez, G., Caballero, J. A., and Jimenez, L. (2008). Application of Life Cycle Assessment to the structural optimization of process flowsheets. *Ind. Eng. Chem. Res.*, 47(3):777 – 789.

Gupta, A., Saini, R., and Sharma, M. (2011). Modelling of hybrid energy system - Part I: Problem formulation and model development. *Renew. Energ.*, 36(2):459 – 465.

Haimes, Y., Lasdon, L., and Wismer, D. (1971). On a bicriterion formulation of the problems of integrated system identification and system optimization. *IEEE Trans. Syst. Man Cybern.*, 1(3):296 –297.

Hajela, P. and Lin, C.-Y. (1992). Genetic search strategies in multicriterion optimal design. *Struct. Optim.*, 4:99 – 107.

Halasz, L., Eder, M., Sandor, N., Niemetz, N., Kettl, K.-H., Tomo, H., and Narodoslawsky, M. (2010). Optimal integration of sustainable technologies in industrial parks. *Chemical Engineering Transactions*, 19:43 – 48.

Han, C. and Stephanopoulos, G. (1996). Knowledge-based approaches in process synthesis. In *International Conference on Intelligent Systems in Process Engineering*, volume 92 of *AIChE Symposium Series*, pages 148 – 159.

Hartono, B., Heidebrecht, P., and Sundmacher, K. (2012). Combined branch and bound method and exergy analysis for energy system design. *Ind. Eng. Chem. Res.*, 51(44):14428 – 14437.

Heckl, I., Friedler, F., and Fan, L. (2010). Solution of separation-network synthesis problems by the P-graph methodology. *Comput. Chem. Eng.*, 34(5):700 – 706.

Hillermeier, C., Hüster, S., Märker, W., and Sturm, T. (2000). Optimisation of power plant design: stochastic and adaptive solution concepts. In Parmee, I. C., editor, *Evolutionary design and manufacture: selected papers from ACDM'00*, pages 3 – 18. Springer-Verlag, Berlin/Heidelberg, 1 edition.

Hittinger, E., Whitacre, J., and Apt, J. (2012). What properties of grid energy storage are most valuable? *J. Power Sourc.*, 206(0):436 – 449.

Horst, R. and Tuy, H. (2010). *Global optimization: deterministic approaches.* Springer-Verlag, Berlin/Heidelberg, 3 edition.

Hui, C.-W. and Natori, Y. (1996). An industrial application using mixed-integer programming technique: A multi-period utility system model. *Comput. Chem. Eng.*, 20, Supplement 2(0):S1577 – S1582.

Huijbregts, M. A. J., Hellweg, S., Frischknecht, R., Hendriks, H. W. M., Hungerbühler, K., and Hendriks, A. J. (2010). Cumulative energy demand as predictor for the environmental burden of commodity production. *Environ. Sci. Tech.*, 44(6):2189 – 2196.

Hwang, C.-L. and Masud, A. S. (1979). *Multiple objective decision making, methods and applications: A state-of-the-art survey.* Springer-Verlag, Berlin/Heidelberg.

IBM ILOG (2011). IBM ILOG CPLEX Optimization studio (version 12.2), User's manual.

Ishibuchi, H. and Murata, T. (1998). A multi-objective genetic local search algorithm and its application to flowshop scheduling. *IEEE Trans. Syst. Man Cybern., Part C: Applications and Reviews*, 28(3):392 – 403.

Iyer, R. R. and Grossmann, I. E. (1997). Optimal multiperiod operational planning for utility systems. *Comput. Chem. Eng.*, 21(8):787 – 800.

Iyer, R. R. and Grossmann, I. E. (1998). Synthesis and operational planning of utility systems for multiperiod operation. *Comput. Chem. Eng.*, 22(7-8):979 – 993.

Jaluria, Y. (2008). *Design and Optimization of Thermal Systems*. McGraw-Hill, New York, USA, 2 edition.

Jennings, M. G., Shah, N., and Fisk, D. (2012). Optimising the arrangement of finance towards large scale refurbishment of housing stock using mathematical programming and optimisationg. In Desideri, U., Manfrida, G., and Sciubba, E., editors, *Proceedings of ECOS 2012: 25th International Conference on Efficiency, Cost, Optimization, Simulation, and Environmental Impact of Energy Systems*, pages 127: 1 – 16, Perugia, Italy.

Jiayi, H., Chuanwen, J., and Rong, X. (2008). A review on distributed energy resources and microgrid. *Renew. Sustain. Energ. Rev.*, 12(9):2472 – 2483.

Juul, N. and Meibom, P. (2011). Optimal configuration of an integrated power and transport system. *Energy*, 36(5):3523 – 3530.

Kaldellis, J. and Zafirakis, D. (2007). Optimum energy storage techniques for the improvement of renewable energy sources-based electricity generation economic efficiency. *Energy*, 32(12):2295 – 2305.

Kallrath, J. (2000). Mixed integer optimization in the chemical process industry: Experience, potential and future perspectives. *Chem. Eng. Res. Des.*, 78(6):809 – 822.

Kallrath, J. and Wilson, J. M. (1997). *Business optimisation using mathematical programming*. MacMillan, London.

Kang, C. A., Brandt, A. R., and Durlofsky, L. J. (2011). Optimal operation of an integrated energy system including fossil fuel power generation, CO_2 capture and wind. *Energy*, 36(12):6806 – 6820.

Kasas, M., Kravanja, Z., and Pintaric, Z. N. (2011). Suitable modeling for process flow sheet optimization using the correct economic criterion. *Ind. Eng. Chem. Res.*, 50(6):3356 – 3370.

Kavvadias, K. and Maroulis, Z. (2010). Multi-objective optimization of a trigeneration plant. *Energ. Pol.*, 38(2):945 – 954.

Keirstead, J., Samsatli, N., Shah, N., and Weber, C. (2012). The impact of CHP (combined heat and power) planning restrictions on the efficiency of urban energy systems. *Energy*, 41(1):93 – 103.

Keirstead, J. and Shah, N. (2011). Calculating minimum energy urban layouts with mathematical programming and Monte Carlo analysis techniques. *Comput. Environ. Urban Syst.*, 35(5):368 – 377.

Kirkpatrick, S., Gelatt, C. D., and Vecchi, M. P. (1983). Optimization by simulated annealing. *Science*, 220(4598):671 – 680.

Kirkwood, R., Locke, M., and Douglas, J. (1988). A prototype expert system for synthesizing chemical process flowsheets. *Comput. Chem. Eng.*, 12(4):329 – 343.

Klatt, K.-U. and Marquardt, W. (2009). Perspectives for process systems engineering - personal views from academia and industry. *Comput. Chem. Eng.*, 33(3):536 – 550.

Klemes, J. J., Friedler, F., Bulatov, I., and Varbanov, P. S. (2011). *Sustainability in the process industry: integration and optimization.* McGraw Hill, New York City.

Koch, C., Cziesla, F., and Tsatsaronis, G. (2007). Optimization of combined cycle power plants using evolutionary algorithms. *Chem. Eng. Process: Process Intensification*, 46(11):1151 – 1159.

Kott, A. S., May, J. H., and Hwang, C. C. (1989). An autonomous artificial designer of thermal energy systems. Part 1 - Theoretical considerations. *J. Eng. Gas Turb. Power*, 111(4):728 – 733.

Kovács, Z., Ercsey, Z., Friedler, F., and Fan, L. T. (2000). Separation-network synthesis: global optimum through rigorous super-structure. *Comput. Chem. Eng.*, 24(8):1881 – 1900.

Krajacic, G., Duic, N., and da Graça Carvalho, M. (2009). H2RES, Energy planning tool for island energy systems - The case of the Island of Mljet. *Int. J. Hydrogen Energ.*, 34(16):7015 – 7026.

Kravanja, Z. (2010). Challenges in sustainable integrated process synthesis and the capabilities of an MINLP process synthesizer MipSyn. *Comput. Chem. Eng.*, 34(11):1831 – 1848.

Kravanja, Z. and Grossmann, I. E. (1993). Prosyn – an automated topology and parameter process synthesizer. *Comput. Chem. Eng.*, 17(Supplement 1):S87 – S94.

Kusiak, A., Tang, F., and Xu, G. (2011a). Multi-objective optimization of HVAC system with an evolutionary computation algorithm. *Energy*, 36(5):2440 – 2449.

Kusiak, A., Xu, G., and Tang, F. (2011b). Optimization of an HVAC system with a strength multi-objective particle-swarm algorithm. *Energy*, 36(10):5935 – 5943.

Lam, H. L., Klemes, J. J. J., Kravanja, Z., and Varbanov, P. S. (2011). Software tools overview: process integration, modelling and optimisation for energy saving and pollution reduction. *Asia Pac. J. Chem. Eng.*, 6(5):696 – 712.

Lam, H. L., Varbanov, P. S., and Klemes, J. J. (2010). Optimisation of regional energy supply chains utilising renewables: P-graph approach. *Comput. Chem. Eng.*, 34(5):782 – 792.

Land, A. H. and Doig, A. G. (1960). An automatic method of solving discrete programming problems. *Econometrica*, 28(3):497 – 520.

Lewin, D. R., Wang, H., and Shalev, O. (1998). A generalized method for HEN synthesis using stochastic optimization - I. General framework and MER optimal synthesis. *Comput. Chem. Eng.*, 22(10):1503 – 1513.

Li, X. and Kraslawski, A. (2004). Conceptual process synthesis: past and current trends. *Chem. Eng. Process*, 43(5):583 – 594.

Lin, Y.-C., Fan, L. T., Shafie, S., Bertok, B., and Friedler, F. (2009). Generation of light hydrocarbons through Fischer-Tropsch synthesis: Identification of potentially dominant catalytic pathways via the graph-theoretic method and energetic analysis. *Comput. Chem. Eng.*, 33(6):1182 – 1186.

Linnhoff, B., Townsend, D. W., Boland, D., Hewitt, G. F., Thomas, B. E. A., Guy, A. R., and Marsland, R. H. (1982). *A user guide on process integration for the efficient use of energy*. The Institution of Chemical Engineers, Rugby, UK.

Liu, P., Georgiadis, M. C., and Pistikopoulos, E. N. (2011). Advances in energy systems engineering. *Ind. Eng. Chem. Res.*, 50(9):4915 – 4926.

Liu, P., Pistikopoulos, E. N., and Li, Z. (2010a). An energy systems engineering approach to the optimal design of energy systems in commercial buildings. *Energ. Pol.*, 38(8):4224 – 4231.

Liu, P., Pistikopoulos, E. N., and Li, Z. (2010b). A multi-objective optimization approach to polygeneration energy systems design. *AIChE J.*, 56(5):1218 – 1234.

Liu, Y.-H. (2010). Different initial solution generators in genetic algorithms for solving the probabilistic traveling salesman problem. *Appl. Math. Comput.*, 216(1):125 – 137.

Lozano, M. A., Ramos, J. C., Carvalho, M., and Serra, L. M. (2009). Structure optimization of energy supply systems in tertiary sector buildings. *Energ. Build.*, 41(10):1063 – 1075.

Lu, M. and Motard, R. (1985). Computer-aided total flowsheet synthesis. *Comput. Chem. Eng.*, 9(5):431 – 445.

Lund, H. (2005). Large-scale integration of wind power into different energy systems. *Energy*, 30(13):2402 – 2412.

Luo, X., Wen, Q.-Y., and Fieg, G. (2009). A hybrid genetic algorithm for synthesis of heat exchanger networks. *Comput. Chem. Eng.*, 33(6):1169 – 1181.

Luo, X., Zhang, B., Chen, Y., and Mo, S. (2011). Modeling and optimization of a utility system containing multiple extractions steam turbines. *Energy*, 36(5):3501 – 3512.

Luo, X., Zhang, B., Chen, Y., and Mo, S. (2012). Operational planning optimization of multiple interconnected steam power plants considering environmental costs. *Energy*, 37(1):549 – 561.

Maaranen, H., Miettinen, K., and Penttinen, A. (2007). On initial populations of a genetic algorithm for continuous optimization problems. *J. Global Optim.*, 37(3):405 – 436.

Mahalec, V. and Motard, R. (1977). Evolutionary search for an optimal limiting process flowsheet. *Comput. Chem. Eng.*, 1(2):133 – 147.

Manassaldi, J. I., Mussati, S. F., and Scenna, N. J. (2011). Optimal synthesis and design of heat recovery steam generation (HRSG) via mathematical programming. *Energy*, 36(1):475 – 485.

Manner, R., Mahfoud, S., and Mahfoud, S. W. (1992). Crowding and preselection revisited. In *Parallel Problem Solving From Nature*, pages 27 – 36. North-Holland.

Maréchal, F., Weber, C., and Favrat, D. (2008). Multiobjective design and optimization of urban energy systems. In Pistikopoulos, E. N., Georgiadis, M. C., Kikkinides, E. S., and Dua, V., editors, *Energy systems engineering*, volume 5 of *Process systems engineering*. Wiley-VCH, Weinheim, Germany.

Marquardt, W., Kossack, S., and Kraemer, K. (2008). A framework for the systematic design of hybrid separation processes. *Chin. J. Chem. Eng.*, 16(3):333 – 342.

Mavromatis, S. and Kokossis, A. (1998). Conceptual optimisation of utility networks for operational variations - I. Targets and level optimisation. *Chem. Eng. Sci.*, 53(8):1585 – 1608.

Mavrotas, G. (2009). Effective implementation of the epsilon-constraint method in multiobjective mathematical programming problems. *Appl. Math. Comput.*, 21(3):455 – 465.

Mavrotas, G., Diakoulaki, D., Florios, K., and Georgiou, P. (2008). A mathematical programming framework for energy planning in services' sector buildings under uncertainty in load demand: The case of a hospital in Athens. *Energ. Pol.*, 36(7):2415 – 2429.

McCormick, G. P. (1976). Computability of global solutions to factorable nonconvex programs: Part I - Convex underestimating problems. *Math. Program.*, 10:147 – 175.

Mehleri, E. D., Sarimveis, H., Markatos, N. C., and Papageorgiou, L. G. (2012). A mathematical programming approach for optimal design of distributed energy systems at the neighbourhood level. *Energy*, 44(1):96 – 104.

Melli, R. and Sciubba, E. (1997). A prototype expert system for the conceptual synthesis of thermal processes. *Energ. Convers. Manag.*, 38(15-17):1737 – 1749.

Miettinen, K. (1999). *Nonlinear multiobjective optimization*, volume 12 of *International Series in Operations Research and Management Science*. Kluwer Academic Publishers, Dordrecht.

Misener, R. and Floudas, C. (2012). GloMIQO: Global mixed-integer quadratic optimizer. *J. Global. Optim.*, pages 1 – 48.

Misener, R., Gounaris, C. E., and Floudas, C. A. (2009). Global optimization of gas lifting operations: A comparative study of piecewise linear formulations. *Ind. Eng. Chem. Res.*, 48(13):6098 – 6104.

Mizsey, P. and Fonyo, Z. (1990a). A predictor-based bounding strategy for synthesizing energy integrated total flowsheets. *Comput. Chem. Eng.*, 14(11):1303 – 1310.

Mizsey, P. and Fonyo, Z. (1990b). Toward a more realistic overall process synthesis - The combined approach. *Comput. Chem. Eng.*, 14(11):1213 – 1236.

Morar, M. and Agachi, P. S. (2010). Review: Important contributions in development and improvement of the heat integration techniques. *Comput. Chem. Eng.*, 34(8):1171 – 1179.

Morrison, R. W. and Jong, K. A. D. (2002). Measurement of population diversity. In Collet, P., Fonlupt, C., Hao, J.-K., Lutton, E., and Schoenauer, M., editors, *Artificial Evolution*, volume 2310 of *Lecture Notes in Computer Science*, pages 31 – 41. Springer.

Mostow, J. (1985). Toward better models of the design process. *AI Mag.*, 6(1):44 – 57.

Nakata, T., Silva, D., and Rodionov, M. (2011). Application of energy system models for designing a low-carbon society. *Progr. Energ. Combust. Sci.*, 37(4):462 – 502.

Needham, M. T. (2011). A psychological approach to a thriving resilient community. *Int. J. Bus. Human. Tech.*, 1(3):279 – 283.

Niknam, T., Narimani, M. R., Jabbari, M., and Malekpour, A. R. (2011). A modified shuffle frog leaping algorithm for multi-objective optimal power flow. *Energy*, 36(11):6420 – 6432.

Nishida, N., Stephanopoulos, G., and Westerberg, A. (1981). A review of process synthesis. *AIChE J.*, 27(3):321 – 351.

Nishio, M. and Johnson, A. I. (1979). Strategy for energy system expansion. *Chem. Eng. Progr.*, 73:75.

Nocedal, J. and Wright, S. J. (2000). *Numerical Optimization*. Springer-Verlag.

Ortiga, J., Bruno, J., and Coronas, A. (2011). Selection of typical days for the characterisation of energy demand in cogeneration and trigeneration optimisation models for buildings. *Energ. Convers. Manag.*, 52(4):1934 – 1942.

Østergaard, P. A. (2009). Reviewing optimisation criteria for energy systems analyses of renewable energy integration. *Energy*, 34(9):1236 – 1245.

Pajula, E., Seuranen, T., Koiranen, T., and Hurme, M. (2001). Synthesis of separation processes by using case-based reasoning. *Comput. Chem. Eng.*, 25(4-6):775 – 782.

Papalexandri, K. P., Pistikopoulos, E. N., and Kalitventzeff, B. (1998). Modelling and optimization aspects in energy management and plant operation with variable energy demands-application to industrial problems. *Comput. Chem. Eng.*, 22(9):1319 – 1333.

Papoulias, S. A. and Grossmann, I. E. (1983). A structural optimization approach in process synthesis - I: Utility systems. *Comput. Chem. Eng.*, 7(6):695 – 706.

Patel, B., Hildebrandt, D., and Glasser, D. (2009). Process synthesis targets: a new approach to teaching design. In El-Halwagi, M. M. and Linninger, A. A., editors, *Design for energy and the environment. Proceedings of the Seventh International Conference on the Foundations of Computer-Aided Process Design*, pages 699 – 708. CRC Press.

Peters, M. S., Timmerhaus, K., and West, R. E. (2003). *Plant design and economics for chemical engineers*. McGraw-Hill, New York, USA, 5th edition.

Petersen, C. C. (1971). A note on transforming the product of variables to linear form in linear programs. Working Paper, Purdue University.

Pezzini, P., Gomis-Bellmunt, O., and Sudrià-Andreu, A. (2011). Optimization techniques to improve energy efficiency in power systems. *Renew. Sustain. Energ. Rev.*, 15(4):2028 – 2041.

Piacentino, A. and Cardona, F. (2008). EABOT - Energetic analysis as a basis for robust optimization of trigeneration systems by linear programming. *Energ. Convers. Manag.*, 49(11):3006 – 3016.

Pillai, J., Heussen, K., and Østergaard, P. (2011). Comparative analysis of hourly and dynamic power balancing models for validating future energy scenarios. *Energy*, 36(5):3233 – 3243.

Pohekar, S. and Ramachandran, M. (2004). Application of multi-criteria decision making to sustainable energy planning - a review. *Renew. Sustain. Energ. Rev.*, 8(4):365 – 381.

Prokopakis, G. J. and Maroulis, Z. B. (1996). Real-time management and optimisation of industrial utilities systems. *Comput. Chem. Eng.*, 20, Supplement 1(0):S623 – S628.

Psarris, P. and Floudas, C. A. (1990). Improving dynamic operability in MIMO systems with time delays. *Chem. Eng. Sci.*, 45(12):3505 – 3524.

Puchinger, J. and Raidl, G. R. (2005). Combining metaheuristics and exact algorithms in combinatorial optimization: a survey and classification. In *Proceedings of the first international work-conference on the interplay between natural and artificial computation conference on Artificial Intelligence and Knowledge Engineering Applications: a bioinspired approach*, volume 2 of *IWINAC'05*, pages 41 – 53, Berlin, Heidelberg. Springer-Verlag.

Raidl, G. R. (2006). A unified view on hybrid metaheuristics. In Almeida, F., Aguilera, M. J. B., Blum, C., Moreno-Vega, J. M., Pérez, M. P., Roli, A., and Sampels, M., editors, *Hybrid Metaheuristics*, volume 4030 of *Lecture Notes in Computer Science*, pages 1 – 12. Springer.

Ralphs, T. K. and Güzelsoy, M. (2006). Duality and warm starting in integer programming. In *Proceedings of the 2006 NSF Design, Service, and Manufacturing Grantees and Research Conference*.

Raman, R. and Grossmann, I. (1991). Relation between MILP modelling and logical inference for chemical process synthesis. *Comput. Chem. Eng.*, 15(2):73 – 84.

Rebennack, S., Kallrath, J., and Pardalos, P. M. (2011). Optimal storage design for a multi-product plant: A non-convex MINLP formulation. *Comput. Chem. Eng.*, 35(2):255 – 271.

Rhinehart, R. R., Su, M., and Manimegalai-Sridhar, U. (2012). Leapfrogging and synoptic leapfrogging: a new optimization approach. *Comput. Chem. Eng.*, 40(0):67 – 81.

Röhrlich, M., Mistry, M., Martens, P., Buntenbach, S., Ruhrberg, M., Dienhart, M., Briem, S., Quinkertz, R., Alkan, Z., and Kugeler, K. (2000). A method to calculate the cumulative energy demand (CED) of lignite extraction. *Int. J. Life Cycle Assess.*, 5:369 – 373.

Ribeiro, P., Johnson, B., Crow, M., Arsoy, A., and Liu, Y. (2001). Energy storage systems for advanced power applications. *Proc. IEEE*, 89(12):1744 –1756.

Rodriguez, M. A. and Vecchietti, A. (2013). A comparative assessment of linearization methods for bilinear models. *Comput. Chem. Eng.*, 48(0):218 – 233.

Rogner, H. H. and Zhou, D. (2007). IPCC fourth assessment report (AR4) - Climate change 2007: Mitigation of climate change.

Rong, A., Lahdelma, R., and Luh, P. B. (2008). Lagrangian relaxation based algorithm for trigeneration planning with storages. *Eur. J. Oper. Res.*, 188(1):240 – 257.

Rozenberg, G., editor (1999). *Handbook of graph grammars and computing by graph transformation: Volume I. Foundations*. World Scientific Publishing Co., Inc., River Edge, NJ, USA.

Rudolph, G. (1994). An evolutionary algorithm for integer programming. In *Parallel Problem Solving from Nature*, pages 139 – 148.

Sagastizábal, C. (2012). Divide to conquer: decomposition methods for energy optimization. *Math. Program.*, 134:187 – 222.

Sama, D., Sanhong, Q., and Gaggioli, R. (1989). A common-sense second law approach for improving process efficiencies. In Ruixian, C. and Moran, M., editors, *Proceeding of International Symposium on Thermodynamic analysis and Improvement of Energy Systems*, pages 520 – 531, Beijing, China. Pergamon Press, Oxford.

Sand, G., Till, J., Tometzki, T., Urselmann, M., Engell, S., and Emmerich, M. (2008). Engineered versus standard evolutionary algorithms: A case study in batch scheduling with recourse. *Comput. Chem. Eng.*, 32(11):2706 – 2722.

Schaffer, J. D. (1985). Multiple objective optimization with vector evaluated genetic algorithms. In *Proceedings of the 1st International Conference on Genetic Algorithms*, pages 93 – 100, Hillsdale, NJ, USA. L. Erlbaum Associates Inc.

Schembecker, G. and Simmrock, K. H. (1997). Heuristic-numeric design of separation processes for azeotropic mixtures. *Comput. Chem. Eng.*, 21, Supplement(0):S231 – S236.

Scheunemann, A. and Becker, M. (2004). *Kennziffernkatalog: Investitionsvorbereitung in der Energiewirtschaft*. Energy Consulting, Gesellschaft für Energiemanagement. German.

Sciubba, E. and Melli, R. (1998). *Artificial intelligence in thermal systems design: concepts and applications*. Nova Science Publishers.

Söderman, J. and Pettersson, F. (2006). Structural and operational optimisation of distributed energy systems. *Appl. Therm. Eng.*, 26(13):1400 – 1408.

Semadeni, M. (2004). Storage of energy, overview. In Cleveland, C. J., editor, *Encyclopedia of Energy*, pages 719 – 738. Elsevier, New York.

Siirola, J. J., Powers, G. J., and Rudd, D. F. (1971). Synthesis of system designs: III. Toward a process concept generator. *AIChE J.*, 17(3):677 – 682.

Smith, R. (2005). *Chemical process design and integration*. John Wiley & Sons, Ltd., Chichester, UK, 2 edition.

Soares, J., Silva, M., Sousa, T., Vale, Z., and Morais, H. (2012). Distributed energy resource short-term scheduling using signaled particle swarm optimization. *Energy*, 42(1):466 – 476.

Solomon, S., Qin, D., Manning, M., Chen, Z., Marquis, M., Averyt, K. B., Tignor, M., and Miller, H. L. (2007). *IPCC Fourth Assessment Report: Climate Change 2007, The Physical Science Basis*. Cambridge University Press.

Spall, J. C. (2003). *Introduction to stochastic search and optimization: estimation, simulation, and control*. John Wiley & Sons, Ltd., Hoboken, New Jersey.

Stahl, T. and Völter, M. (2006). *Model-driven software development: technology, engineering, management*. John Wiley & Sons, Ltd., Chichester, UK.

Stefansson, H., Sigmarsdottir, S., Jensson, P., and Shah, N. (2011). Discrete and continuous time representations and mathematical models for large production scheduling problems: A case study from the pharmaceutical industry. *Eur. J. Oper. Res.*, 215(2):383 – 392.

Steuer, R. E. (1986). *Multiple criteria optimization: theory, computation and application*. John Wiley & Sons, Ltd., Hoboken, New Jersey.

Sun, J. and Li, W. (2011). Operation optimization of an organic Rankine cycle (ORC) heat recovery power plant. *Appl. Therm. Eng.*, 31(11-12):2032 – 2041.

Tawarmalani, M. and Sahinidis, N. V. (2004). Global optimization of mixed-integer nonlinear programs: A theoretical and computational study. *Math. Program.*, 99:563 – 591.

Tawarmalani, M. and Sahinidis, N. V. (2005). A polyhedral branch-and-cut approach to global optimization. *Math. Program.*, 103(2):225 – 249.

Tina, G. and Passarello, G. (2012). Short-term scheduling of industrial cogeneration systems for annual revenue maximisation. *Energy*, 42(1):46 – 56.

Toffolo, A. and Lazzaretto, A. (2002). Evolutionary algorithms for multi-objective energetic and economic optimization in thermal system design. *Energy*, 27(6):549 – 567.

Tveit, T.-M., Savola, T., Gebremedhin, A., and Fogelholm, C.-J. (2009). Multi-period MINLP model for optimising operation and structural changes to CHP plants in district heating networks with long-term thermal storage. *Energ. Convers. Manag.*, 50(3):639 – 647.

Uhlenbruck, S. and Lucas, K. (2004). Exergoeconomically-aided evolution strategy applied to a combined cycle power plant. *Int. J. Therm. Sci.*, 43(3):289 – 296.

United Nations (1987). Report of the World Commission on Environment and Development, General Assembly Resolution. A/RES/42/187, 96th plenary meeting.

Urselmann, M., Emmerich, M. T. M., Till, J., Sand, G., and Engell, S. (2007). Design of problem-specific evolutionary algorithm/mixed-integer programming hybrids: two-stage stochastic integer programming applied to chemical batch scheduling. *Eng. Optim.*, 39(5):529 – 549.

Varbanov, P., Doyle, S., and Smith, R. (2004). Modelling and optimization of utility systems. *Chem. Eng. Res. Des.*, 82(5):561 – 578.

Varbanov, P. S., Klemes, J., and Friedler, F. (2011). Integration of fuel cells and renewables into efficient CHP systems. In Bojic, M., Lior, N., Petrovic, J., Stefanovic, G., and Stevanovic, V., editors, *Proceedings of ECOS 2011: 24th International Conference on Efficiency, Cost, Optimization, Simulation, and Environmental Impact of Energy Systems*, pages 1021 – 1033, Novi Sad, Serbia.

Varbanov, P. S., Perry, S., Klemes, J. J., and Smith, R. (2005). Synthesis of industrial utility systems: cost-effective de-carbonisation. *Appl. Therm. Eng.*, 25(7):985 – 1001.

Velasco-Garcia, P., Varbanov, P. S., Arellano-Garcia, H., and Wozny, G. (2011). Utility systems operation: Optimisation-based decision making. *Appl. Therm. Eng.*, 31(16):3196 – 3205.

Virmani, S., Adrian, E., Imhof, K., and Mukherjee, S. (1989). Implementation of a lagrangian relaxation based unit commitment problem. *IEEE Trans. Power Syst.*, 4(4):1373 – 1380.

Voll, P., Kirschbaum, S., and Bardow, A. (2010). Evaluation of quasi-stationary simulation for the analysis of industrial energy systems. In Favrat, D. and Maréchal, F., editors, *Proceedings of ECOS 2010: 23rd International Conference on Efficiency, Cost, Optimization, Simulation and Environmental Impact of Energy Systems*, volume 4, pages 219 – 231, Lausanne, Switzerland. CreateSpace.

Voll, P., Klaffke, C., Hennen, M., and Bardow, A. (2013). Automated superstructure-based synthesis and optimization of distributed energy supply systems. *Energy*, 50(0):374 – 388.

Voll, P., Lampe, M., Wrobel, G., and Bardow, A. (2012). Superstructure-free synthesis and optimization of distributed industrial energy supply systems. *Energy*, 45(1):424 – 435.

Wang, J.-J., Jing, Y.-Y., Zhang, C.-F., and Zhao, J.-H. (2009). Review on multi-criteria decision analysis aid in sustainable energy decision-making. *Renew. Sustain. Energ. Rev.*, 13(9):2263 – 2278.

Wächter, A. and Biegler, L. T. (2006). On the implementation of an interior-point filter line-search algorithm for large-scale nonlinear programming. *Math. Program.*, 106:25 – 57.

Weber, C., Maréchal, F., Favrat, D., and Kraines, S. (2006). Optimization of an SOFC-based decentralized polygeneration system for providing energy services in an office-building in tokyo. *Appl. Therm. Eng.*, 26(13):1409 – 1419.

Weber, C. and Shah, N. (2011). Optimisation based design of a district energy system for an eco-town in the united kingdom. *Energy*, 36(2):1292 – 1308.

Westerberg, A. W. (1991). Process engineering, perspectives in chemical engineering, research and education. In Colton, C. K., editor, *Advances in chemical engineering*, volume 16, pages 499 – 523. Academic Press, Boston, USA.

Westerberg, A. W. (2004). A retrospective on design and process synthesis. *Comput. Chem. Eng.*, 28(4):447 – 458.

Williams, H. P. (1999). *Model Building in Mathematical Programming, 4th Edition*. John Wiley & Sons, Ltd., Hoboken, New Jersey, 4 edition.

Wright, J. A., Zhang, Y., Angelov, P., Buswell, R. A., and Hanby, V. I. (2008a). Evolutionary synthesis of HVAC system configurations: algorithm development. *HVAC & R Research*, 14(1):33 – 55.

Wright, J. A., Zhang, Y., Angelov, P., Hanby, V. I., and Buswell, R. A. (2008b). Evolutionary synthesis of HVAC system configurations: experimental results. *HVAC & R Research*, 14(1):57 – 72.

Wuebbles, D. J. and Jain, A. K. (2001). Concerns about climate change and the role of fossil fuel use. *Fuel Process. Tech.*, 71(1 - 3):99 – 119.

Yokoyama, R., Hasegawa, Y., and Ito, K. (2002). A MILP decomposition approach to large scale optimization in structural design of energy supply systems. *Energ. Convers. Manag.*, 43(6):771 – 790.

Yokoyama, R. and Ito, K. (2006). Optimal design of gas turbine cogeneration plants in consideration of discreteness of equipment capabilities. *J. Eng. Gas Turb. Power*, 128(2):336 – 343.

Yokoyama, R. and Ose, S. (2012). Optimization of energy supply systems in consideration of hierarchical relationship between design and operation. In Desideri, U., Manfrida, G., and Sciubba, E., editors, *Proceedings of ECOS 2012: 25th International Conference on Efficiency, Cost, Optimization, Simulation, and Environmental Impact of Energy Systems*, pages 389: 1 – 13, Perugia, Italy.

Zhou, Z., Liu, P., Li, Z., Pistikopoulos, E. N., and Georgiadis, M. C. (2013). Impacts of equipment off-design characteristics on the optimal design and operation of combined cooling, heating and power systems. *Comput. Chem. Eng.*, 48:40 – 47.

Zitzler, E. and Thiele, L. (1998). Multiobjective optimization using evolutionary algorithms - a comparative case study. In *Conference on Parallel Problem Solving from Nature (PPSN V)*, pages 292 – 301, Amsterdam.